Why was it written?

I believe that everybody can learn how to sketch. For designers, it is important because this is the way they establish inner and outer communication in the design process. My vision is to put a powerful and suitable source of helpful tips and inspiration on every designer's table.

This edition is a result of efforts to teach product design engineering students how to sketch and use these techniques in the design process of form and colour classes. Until now, sketching guides were focused on drawing techniques. They were illustrator-orientated and had nothing to do with the design process or methodology. This guide introduces a new system of strategic sketch use in the design process.

I wanted to make the book material easy to follow by using examples to demonstrate the strategic sketch system. These examples are reconstructed design process drawings of 'iconic' products that everyone would easily recognise and understand.

I want to thank the Institute for Product Design of NTNU for making it possible for me to publish this guide. Particular thanks go to Professor Johannes Sigurjonsson, Erling Rohde and Trond Are Øritsland for their great support. I also want to express gratitude to my students for the good atmosphere and constant source of inspiration, and especially Alex Mitchell for tutoring in English.

Nenad Pavel

3

What's It All About?

Sketching is the traditional way of designing. The process of sketching facilitates the process of design. Design is a skill and a sophisticated psychological activity, which means that design can be learnt only by practice, as with sketching. Due to this, sketching is not only a way to design, but a way to learn how to design.

This guide is devoted to design students and young designers, but also to experienced designers as a reminder, for inspiration and as a motivator. It is meant for every day use, as a reference while designing. This guide approaches design as implementation and manipulation of design knowledge i.e. production processes, materials and marketing through drawing and form giving skills. By learning to use drawing skills strategically throughout the design process, designers will improve themselves, especially in use of methodology, ease of conceptualizing, form giving, and raise their creativity.

New media of expression has placed this traditional skill in a new context. Designers no longer use drawing to present their final solutions. Instead, they use CAD models, rendering and animation. Drawing is presently used more for the designer's inner communication in order to direct one's thoughts in the design process. This has affected the importance of drawing in design. Technique and materials are now less important. Designers use sketch tablets, studio tools and such like. The purpose of sketching is much more important. This is the reason that this guide focuses on the strategic sketch system. Emphasis is put on sketching for creating the product solution, right before the presentation CAD or workshop model stage.

The conceptualization and form giving phase is run differently by designers so please note that it is very individual. The intention of this guide is to summarise and give an overview of how sketching can be used in this phase.

Many designers use foam or clay models. Rough foam or wooden models could be a very useful functional note in the design process and clay is very useful when form giving. Some designers like to use CAD, but the problem with CAD is that most software is not designed to facilitate the design process. In fact, software can influence and change the design in its own way. The exception to this is Alias Wavefront's 'Studio Tools'. This software has a special sketching option to be used with a tablet. Also, some designers find it very useful to design a rough model with 3D software in order to trace the printout when drawing. So I would recommend the use of CAD in order to visualise your design or as a method to make a prototype by use of rapid prototyping. Therefore, this guide has no intention of recommending sketching as the only way to design.

The real intention of this guide is to show how to navigate the design process by sketching and this means the focus is not on sketching techniques but the use of them. The drawings exhibited in this book demonstrate that you can achieve satisfactory results in just a short time with few materials. In fact, all the drawings in the Styles subchapter were drawn only once with limited time and materials in order to show that this is possible.

The guide content is a collection of techniques that are to be used whilst designing. So I strongly recommend that you familiarise yourself with the content before starting to use it. The scheme bellow shows the content organisation.

Sketching Materials	Sketch Elements	Pitching Sketch	Memo Sketch	Concept Sketch	Specifi-cation Sketch

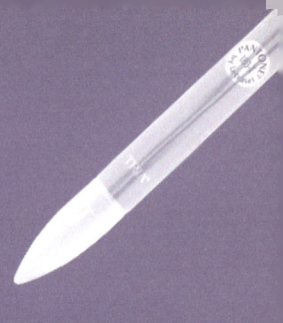

Sketching Materials

markers

pencils and pens

pastels and equipment

papers and equipment

Introduction

Designer's sketching materials differ a lot from that of artists and illustrators. For designer, it is most important to express himself quickly without losing track of the design process. This means that design sketching demands precision and clarity, which depend a lot on the materials used. Wrong material choice or wrong use of materials will act as a barrier between the designer and his idea. By developing your sketching skills, you will develop a sense of materials and the purposes they suit. Until then, please use the materials that are recommended in this guide.

It does not necessarily mean that sketches will be better if you have a great range of materials or use expensive materials. Usually a carefully selected choice of nine markers and five pastels will suffice. This guide shows carefully selected combinations in order to use materials cleverly and save money.

Note
Markers are usually used to imitate clear reflections on the product surface, whilst pastels are used for soft blurred reflections. Coloured pencils and fine liners are used to outline the shape.

According to sketching material properties, best results are achieved if used in this order: markers, pastels, colour pencils, markers refinement if needed.

Markers

- Pantone Cool Gray palette: Cool Gray 2, Cool Gray 5, Cool Gray 8.
- Pantone grey colour palette: 5507-T,452-T,5425-T.
- Pantone colour palette: 180-T,541, 4505-T.
- Non gradient markers: red, blue, black.

Note

   Marker colours must have high monochromatic value, so it is possible to create gradients. Colors such as light green, magenta, yellow or any fluorescent colours are not appropriate for design sketching.

Pencils and Pens

- Propelling pencil 0.5
- Coloured pencils: white, black, blue, green, brown, red
- White chalk pencil
- Fine liners: 0.1, 0.5, 0.9
- Eraser
- Eraser stick

Pastels and Equipment

• Basic pastel palette : black, white, blue, green, red and brown
• Talcum powder
• Lighter fuel
• Cotton wool

Papers and Equipment

• Plain white paper.
• Marker paper.
• Coloured card (colours should be in mid tone)
• Craft Knife
• Masking tape
• Ruler

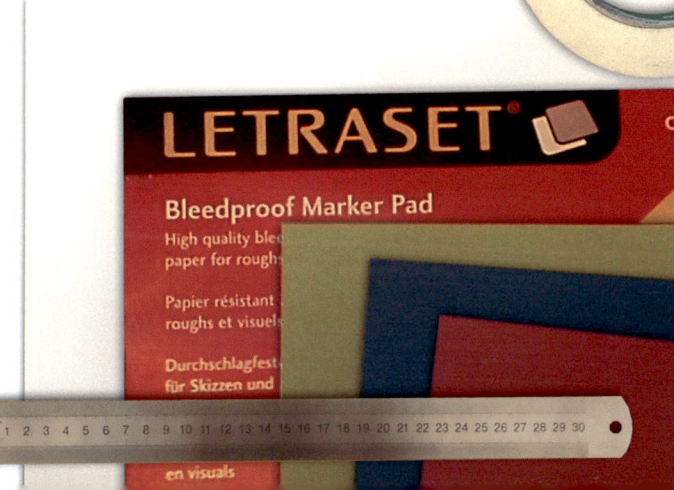

Sketch Elements

sketch layout

object and drawing construction

views

styles

Introduction

-To explain sketching throughout the design process, four elements
are introduced:

• Sketch layout; a content plan.

• Object and drawing construction; shape building.

• Views; describing aspects of the product.

• Styles; techniques to render views.

   Views and styles are combined according to the purpose of the
sketch. For example, if the purpose of the sketch is to make a concept
of the product construction and parts, the required view would be an
explosion in a shaded style. The rest of the guide explains how to use
a combination of views and styles for a certain purposes.

Sketch Layout
A Content Plan

Concept, specification and sometimes pitching sketches need to be presented in some form of a document. This form is used in the design process for communication amongst the design team, and after the process, as a track of its direction. We will call it a sketch layout.

Typical marker paper formats you will face are A3 and A4. Usually the A3 format is the most useful because drawings can be bigger and more detailed. It also leaves more space for designing sketch layouts.

A design sketch always tries to describe design aspects of the product (see the views, page 23). Before starting to fill the paper, it is important to think about what is supposed to be presented. This includes the number, type, size and priority of drawing. Once the content has been decided, it needs to be structured in a logical and readable manner.

Views, text, graphics and backgrounds are used to design a sketch layout. We read from left to right and from top to bottom, so our eyes follow an imaginary diagonal from top left to bottom right. Therefore, it is best to put the title in the top left corner. Sketches form the main content of the layout, which will naturally take the centre of the sheet and text will take the bottom left corner. Backgrounds such as grey squares or frames will help the organisation of the sheet in terms of grouping and emphasising content.

To start your sketch layout, make a small plan of the content no bigger than 5x3 cm. Take care with what is supposed to be presented and explained to the observer and with your use of compositional elements i.e. drawings, graphics, backgrounds and text.

In the case of this showerhead presentation, it was important to present the shape, especially the water jets and rubber component on the top surface. Technical aspects are presented by a ghosting view and text.

Content Plan

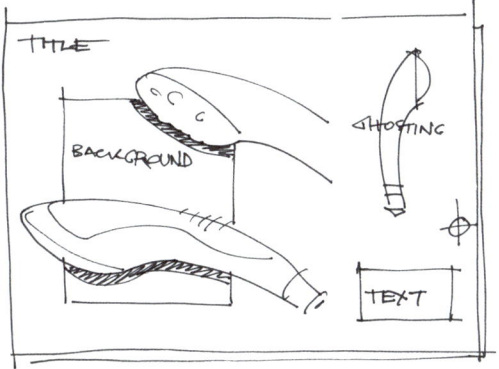

Final Layout

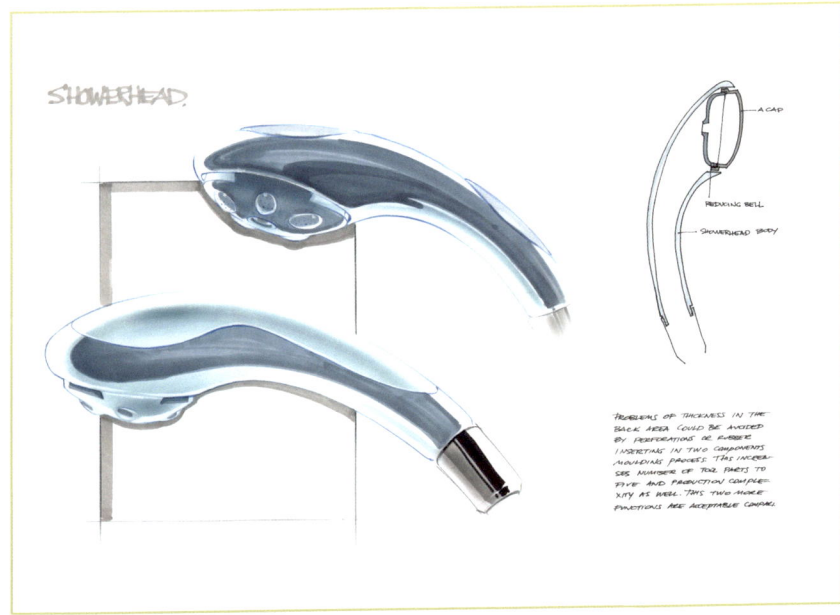

Object and Sketch Construction
Shape Building

Object and drawing construction is used in order to build the shapes you want to present on the sheet.

Note

A design sketch is supposed to describe the product appeal to the user. This means that it slightly exaggerates the form in order to present the product character. For example, if we try to scale the toy car model to the original car size, you would notice that external elements such as handles, mirrors, lights etc, are significantly bigger than on the real car. These elements describe the character of the car, and they must be enlarged to emphasise this. When sketching the product, the appropriate view should explain its size, character and environment.

cubic shapes

revolved shapes

amorphic shapes

Steps:

- Choose a name for your product. Words and branding actually help a lot in the process of getting the product on paper. It helps to describe the function and identity of the product.
- Choose the paper format. Use A4 paper for sketch layouts (see page).This sheet is then very useful during the drawing process because you can put it underneath your hand whilst sketching on the main sheet. This helps to keep your composition in mind and also prevents finger marks.
- Define elements of your product. This step is very important because it will make you aware of the use, shape and construction of the product. Use an A4 sheet to name all the parts and subparts of your product. This is especially important when it comes to complicated interfaces.
- Use the typical projection view. Take care of the product proportion, including the parts and subparts. Use rulers for precision.
- Draw a perspective view. On the following pages, you will see examples of the three product construction types and the steps in order to complete this perspective view.

Cubic Shapes

- Use a propelling pencil to draw perspective lines. Draw the vertical centreline of the monitor. This line defines the symmetry axis. Perspective lines are drawn for the various parts of the product, in this case, the screen and the base. When drawing them, pay attention to the proportion of the parts. Here, the screen represents 2/3 of the product height. The proportions define the shape and character of the product.

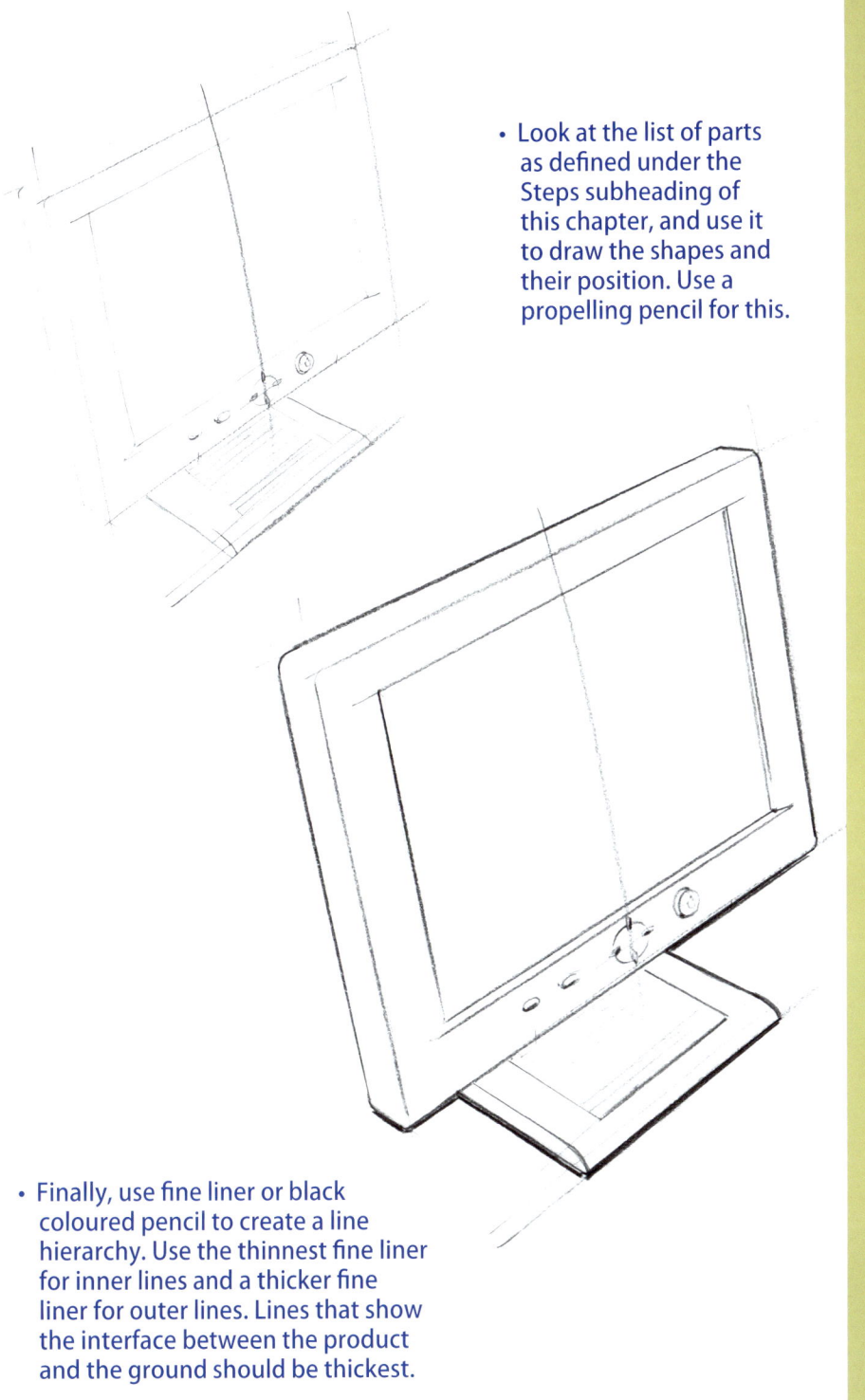

- Look at the list of parts as defined under the Steps subheading of this chapter, and use it to draw the shapes and their position. Use a propelling pencil for this.

- Finally, use fine liner or black coloured pencil to create a line hierarchy. Use the thinnest fine liner for inner lines and a thicker fine liner for outer lines. Lines that show the interface between the product and the ground should be thickest.

Revolved Shapes

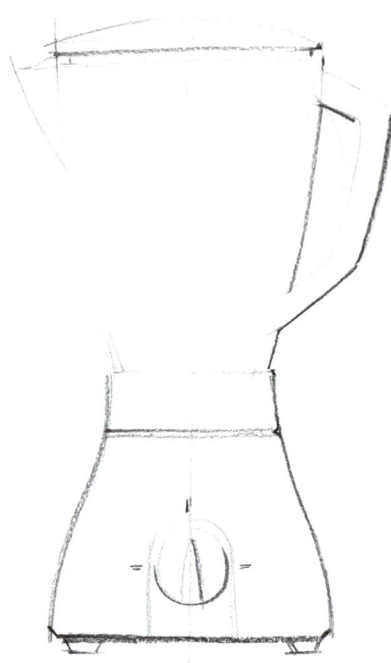

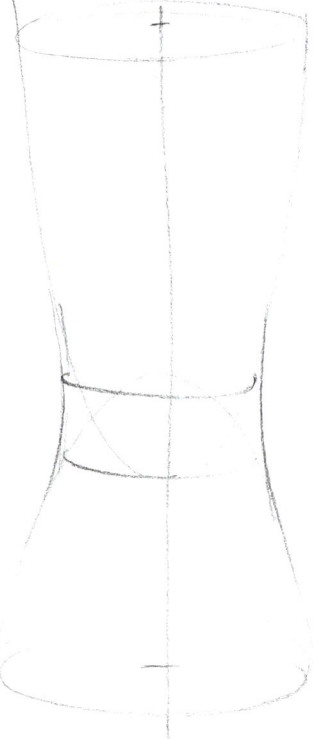

• Use the propelling pencil to draw a vertical centreline. Draw horizontal lines to separate the parts of the blender. Draw vertical perspective lines so that the blender is being viewed from the top. When drawing them, be aware of the proportion of the parts. In this case the blender is narrowest in the middle.

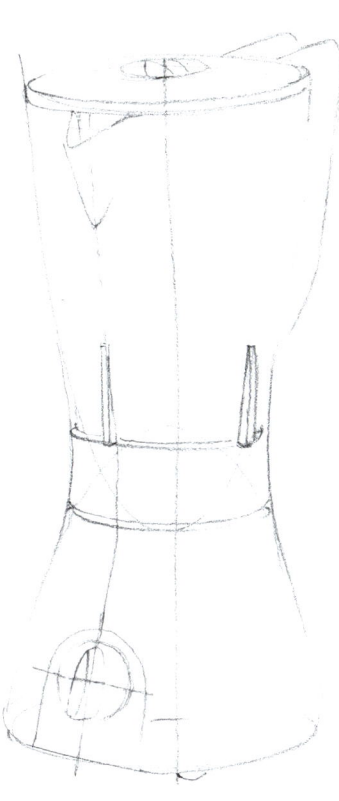

- Look at the list of parts as defined under the Steps subheading of this chapter, and use it to draw the shapes and their position. Use a propelling pencil for this.

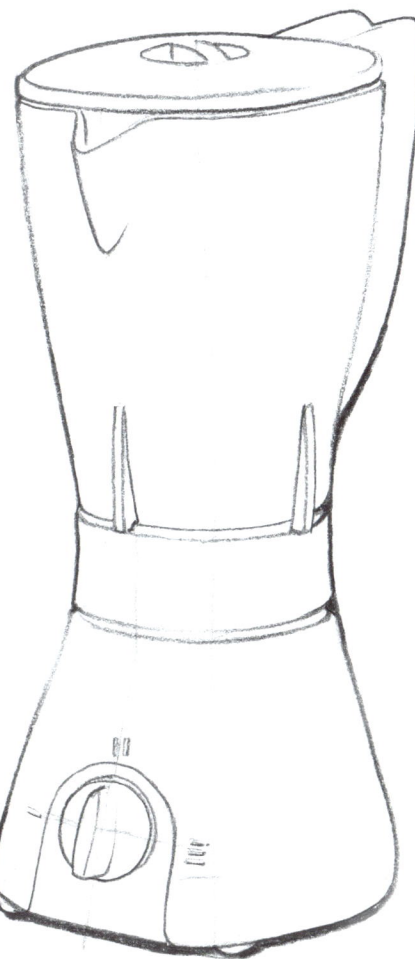

- Finally, use fine liner or black coloured pencil to create a line hierarchy. Use the thinnest fine liner for inner lines and a thicker fine liner for outer lines. Lines that show the interface between the product and the ground should be thickest.

Amorphic Shapes

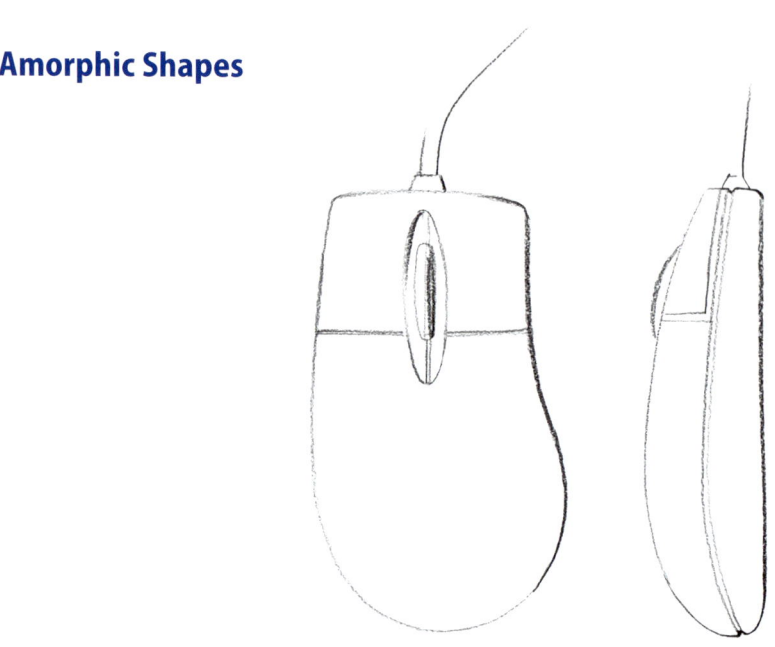

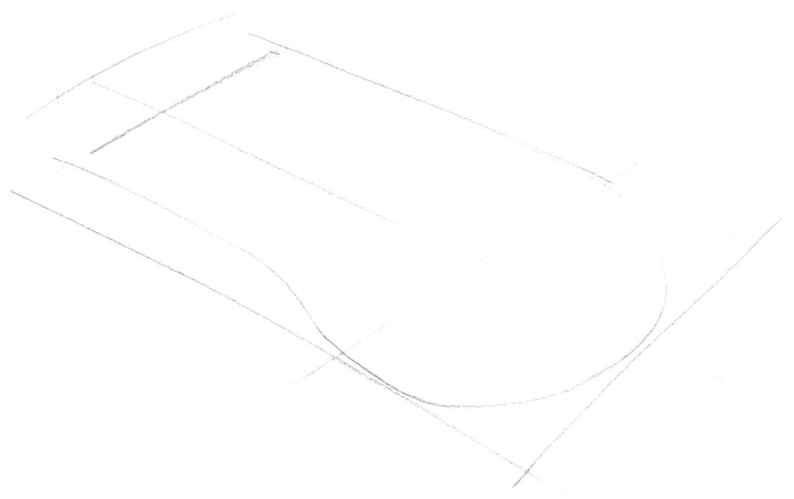

• Construction of amorphic products starts with the base. The easiest way is to draw a rectangle in perspective as the base.

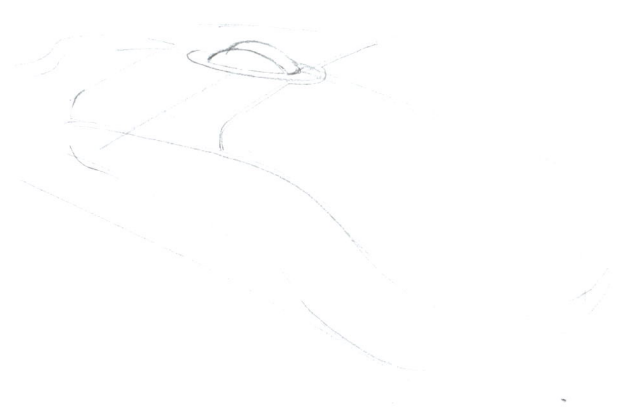

• Define the centreline of the base and also the height of the object.
  Look at the list of parts as defined under the Steps subheading of this
  chapter, and use it to draw the shapes and their position.
  Use a propelling pencil for this.

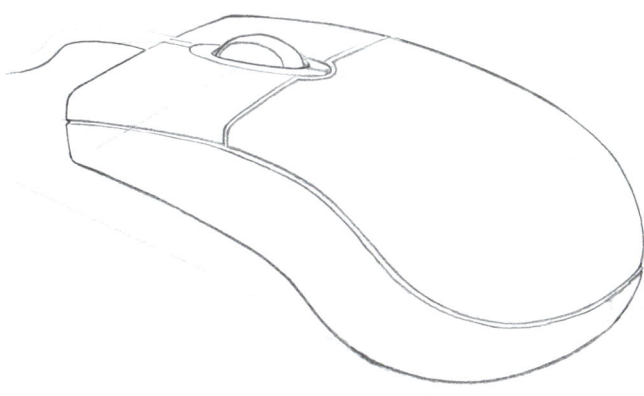

• Finally, use fine liner or black coloured pencil to create a line hierarchy.
  Use the thinnest fine liner for inner lines and a thicker fine liner for
  outer lines. Lines that show the interface between the product and the
  ground should be thickest.

The Views
Describing Aspects of the Product

Views are used as a system to describe different design aspects of the product. Also, different views can be used successfully as a method in the design process and in product presentation.

Projections

Perspective

Netting

Timeline

Star

Explosion

Ghosting

Section

Detail

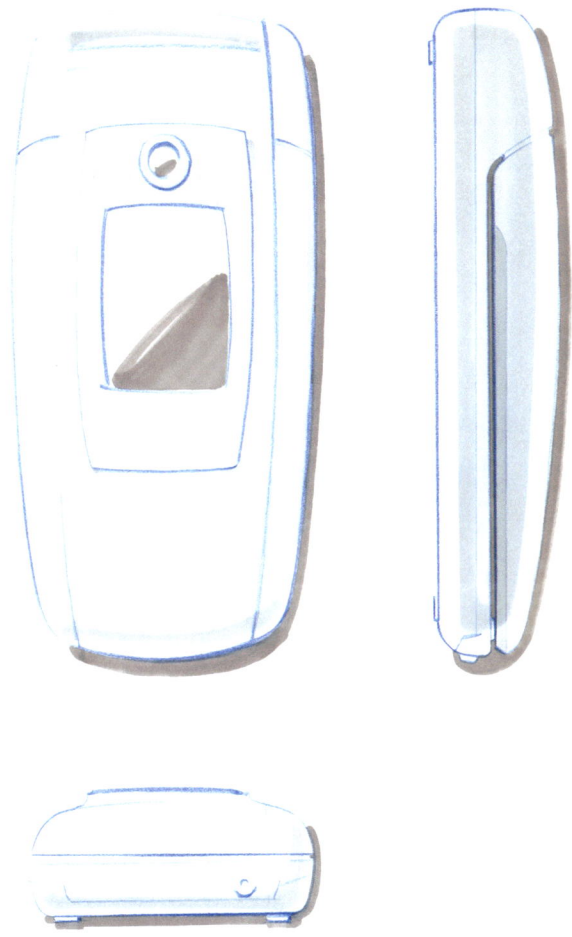

Projections

Projections are Views that describe the product in two dimensions.
It is used for defining the proportions of the product.

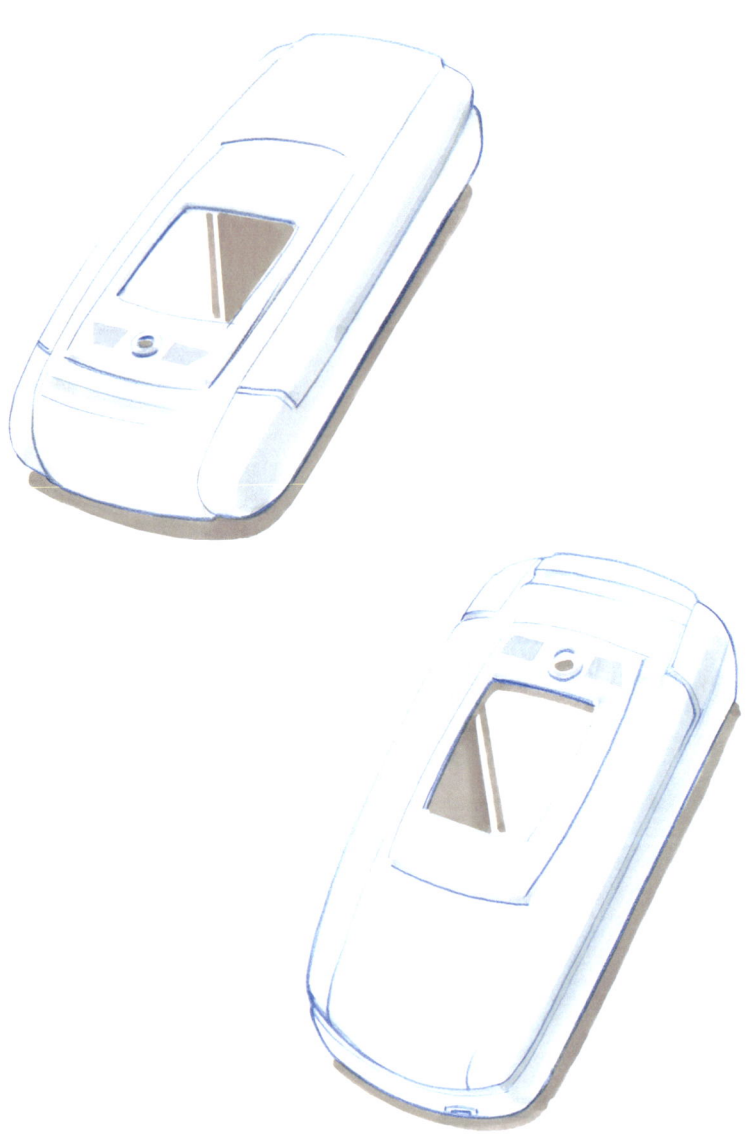

Perspective

This View describes the shape of the product. It is used as an idea generator and as a descriptor.

Netting

A Netting shows lines that describe the form of the product.
It is used for shape specification and detailing.

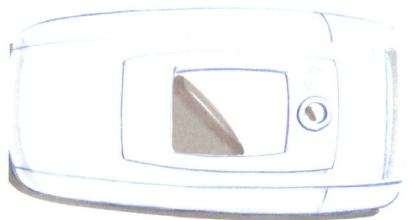

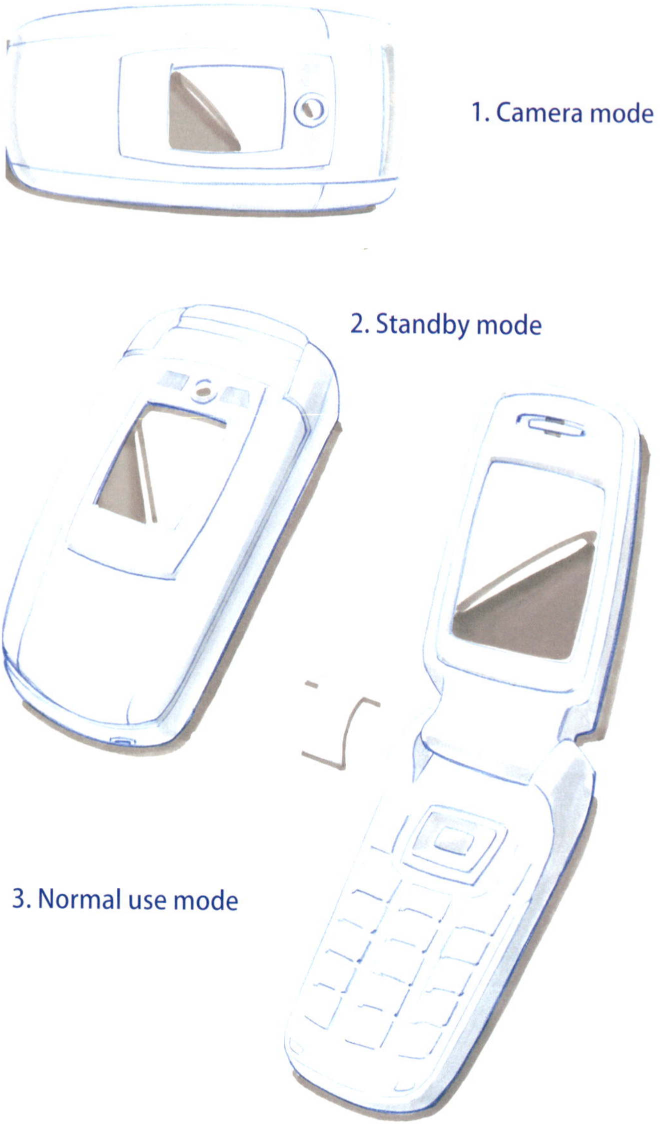

1. Camera mode

2. Standby mode

3. Normal use mode

Time line

This View describes two or more stages in the use of the product. It is used to describe the function and behaviour of the product and can be used to form a scenario.

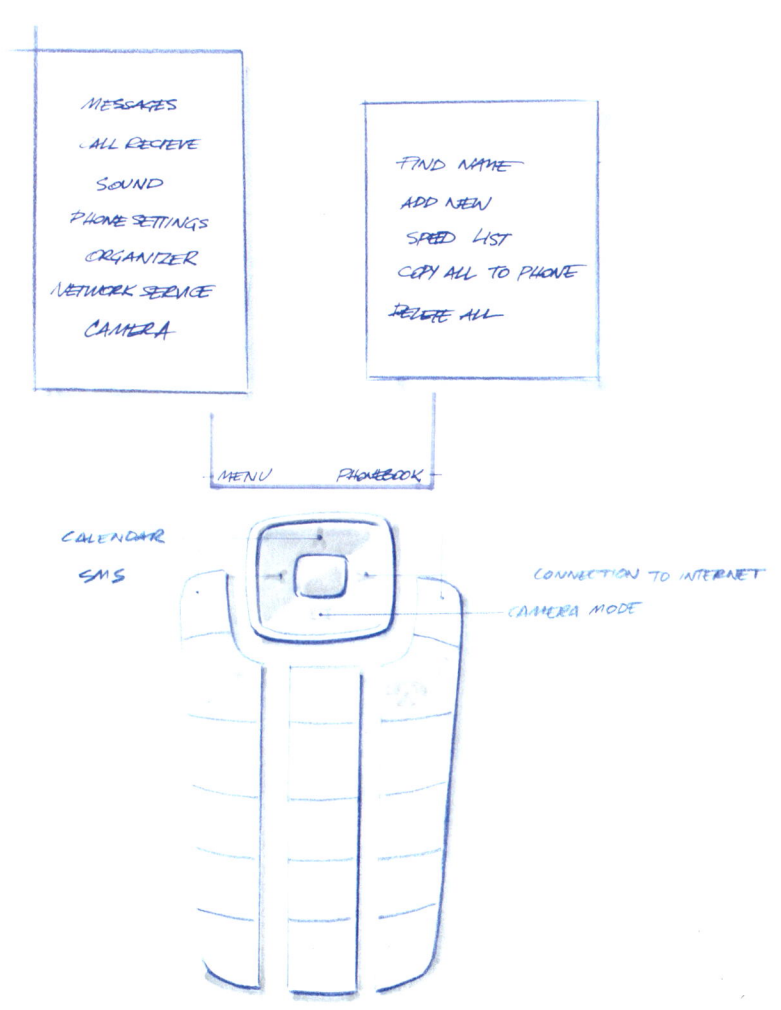

MESSAGES
CALL RECIEVE
SOUND
PHONE SETTINGS
ORGANIZER
NETWORK SERVICE
CAMERA

FIND NAME
ADD NEW
SPEED LIST
COPY ALL TO PHONE
DELETE ALL

MENU   PHONEBOOK

CALENDAR
SMS

CONNECTION TO INTERNET
CAMERA MODE

Star

This View describes the user interface of the product by using 'if and then' explanations.

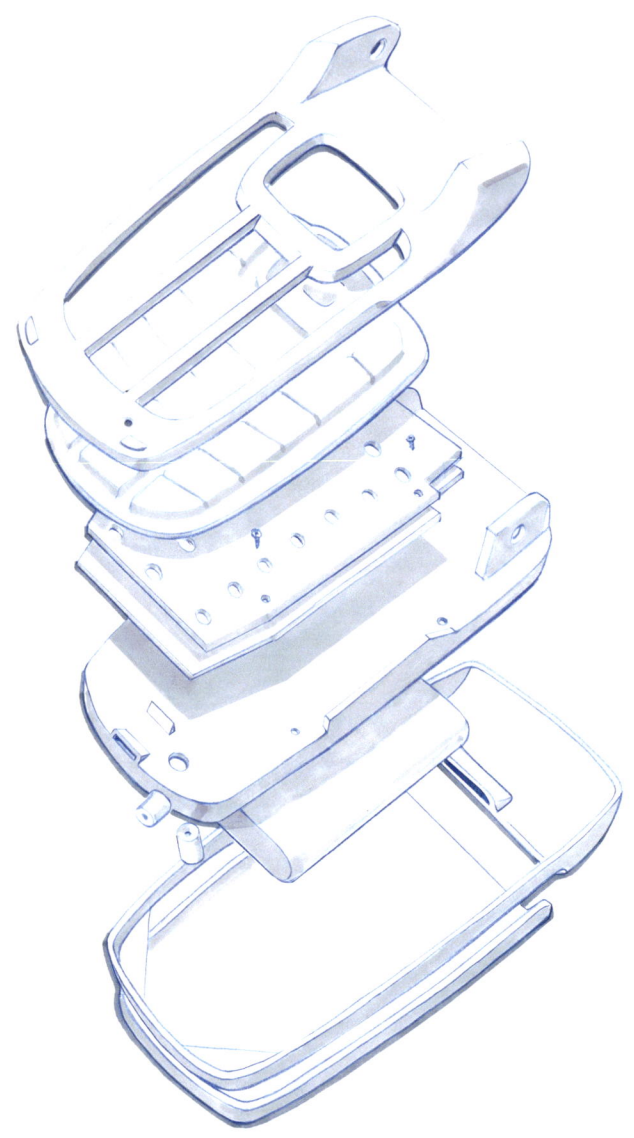

cubic shapespes

Explosion

Explosions show how the parts of the product fit together.

Ghosting

Ghosting describes the shape and position of inner parts by making objects in front of them transparent.

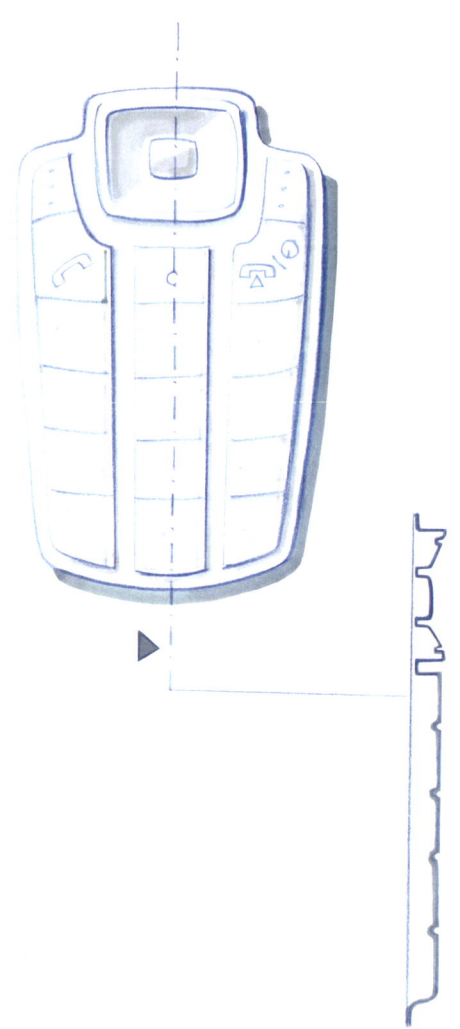

Section

Section Views show a cut-away of the product.

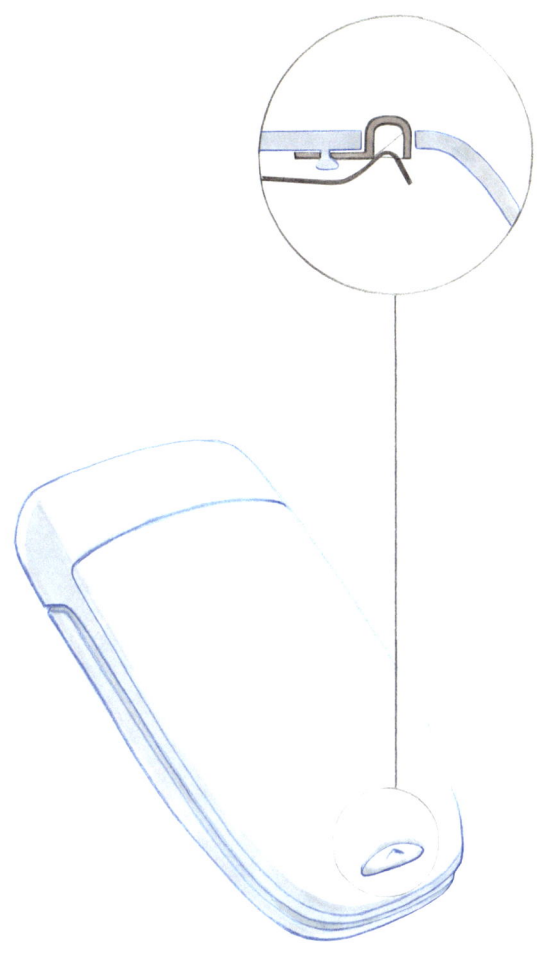

Detail

This is where other Views, as described previously, are magnified in order to describe details more fully.

The Styles
Techniques to Render Views

Light and Shadow

Design drawing is based on shading surfaces of the product. This shading is performed in three tones of one colour. There are two ways to apply and treat colour on the paper, by addition and subtraction. Light tonal areas are always for where light hits the product; dark tones are for shadow or object colour.

The images on the opposite page present the three basic rendering techniques to show how colour and light may affect the surface of the object. If we imagine the light coming from the top left, then the right side of the box will have the darkest tone. The left side will be the next darkest and the top will be lightest. Please notice how the concave and convex domed surfaces are presented. All the convex domed surfaces will reflect light on their left side and the right side will be in shadow. The concave domed surfaces will be the opposite. This principle is also used to create grooves from simple lines. To emphasise light, all the outer lines will be thicker than inner lines. Also, lines on the shaded side will be thicker then those on the light one.

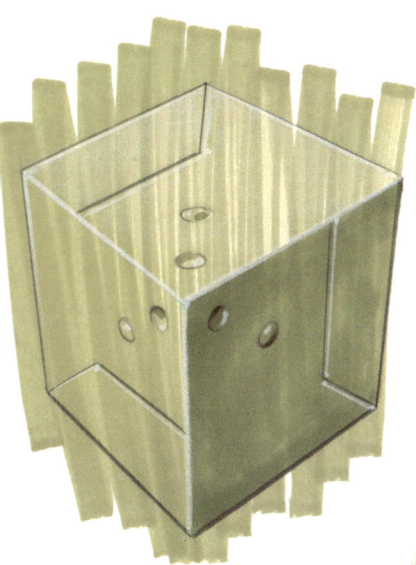

- One layer of marker tone represents the natural colour of the object. Three layers represent shade on the right side, and a layer of white pastel is applied to the top to show light hitting the object.

- The background is created by applying pastel mixed with lighter fuel. This tone represents the natural colour of the object. The sides in the shadow are covered by layer of black coloured pencil and highlights are created by erasing the pastel.

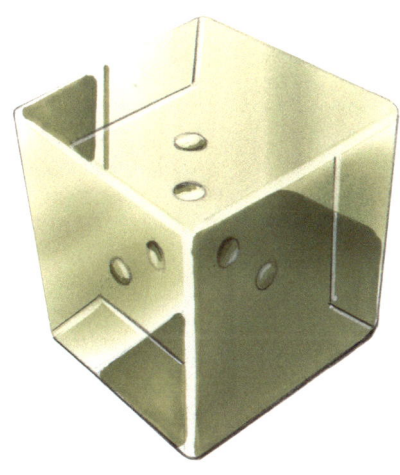

- This is a mixture of the two techniques above. It gives the impression of highly reflective surfaces.

The sketching of Industrial designers is based on clichés. These clichés are used as a method to communicate the shape, texture, material and colour of a product. It is not only the final rendering that is a cliché, but also how the materials are used to produce the rendering. Therefore, the definition of a Style is the skill of applying rendering clichés to Views quickly in order to concentrate on the design process.

free style

outline drawing

shading

highlight

rubber and matt plastic

reflective materials

brushed metal

textures

Free Style

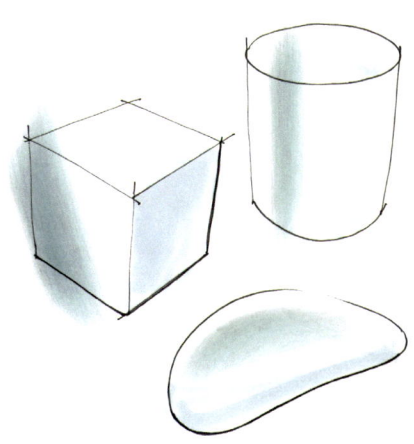

- Free Style is a mixture of all the styles you will find in this book. It is the fastest and easiest way to describe a view.

- Use pantone 5507-T to draw the object shadows.

- Pastel dust mixed with talcum powder is used for soft shadows. Apply it with cotton wool as a gradient from bottom left to top right in straight strokes.

- You can either use black coloured pencil or fine liner for outlining. Inner lines are always thin whilst outer lines are thicker. If using fine liner, use 0.1 thickness for inner lines and 0.9 for outer lines. Eraser sticks (see Sketching Materials) can add highlights and edges reflecting light (see Light and Shadow).

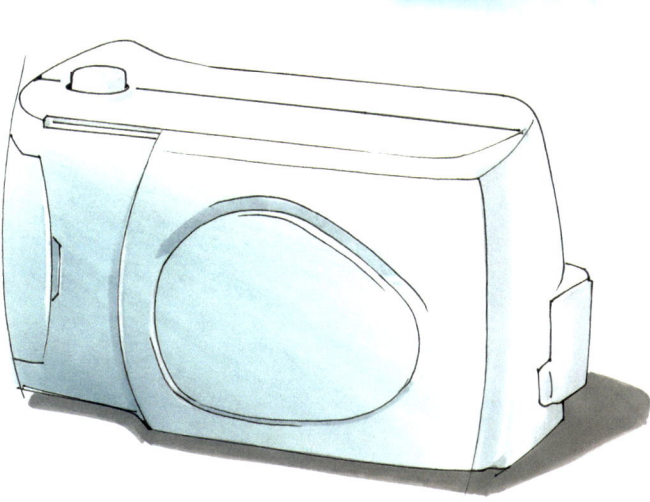

Outline Drawing

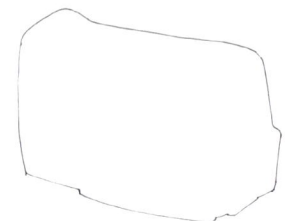

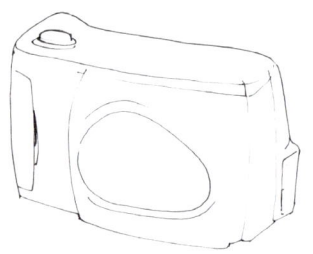

- Use a fine liner to draw an outline silhouette of the product. Do not construct the shape.

- Add product elements to describe the most important parts and their shape.

- Insert shadows in specific areas and use the thick side of a marker to draw the shadow of the object on the ground.

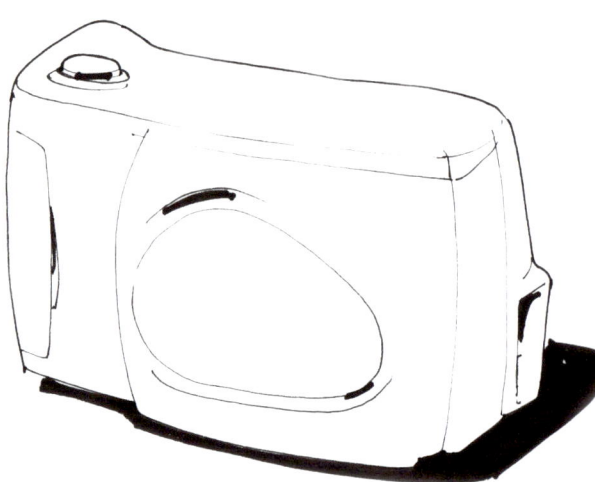

Shading

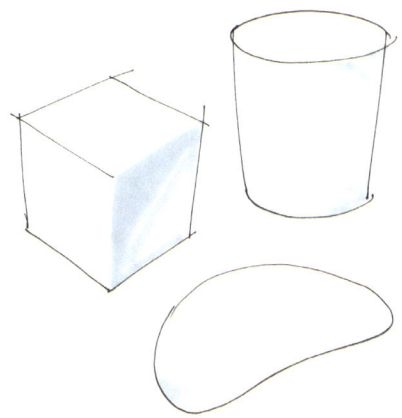

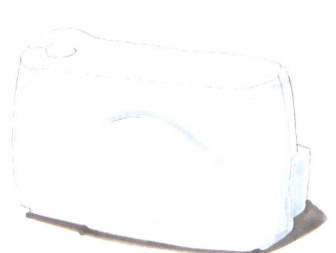

- Use the propelling pencil for Object and Drawing Construction.

- Use Pantone 5507-T to draw shadows on the object and Cool Gray 7 for the shadows that the object casts.

- Use 0.1 fine liner for inner lines and 0.9 for outer lines.

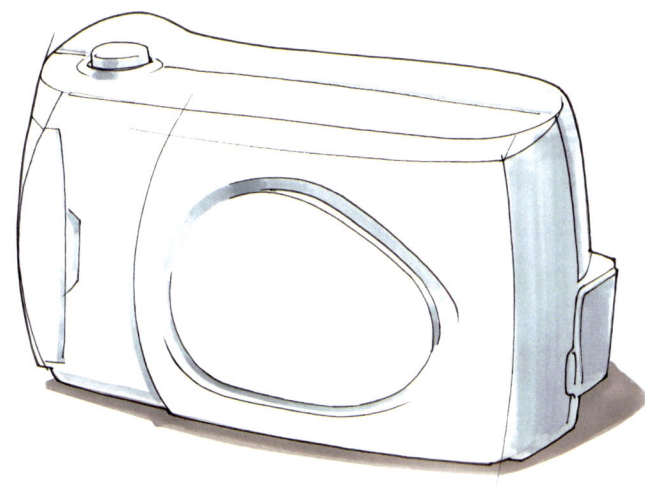

Highlight

- If you choose to use coloured card, go instantly to the next step. If using white marker paper, follow this procedure to make the coloured background for the shape: Grate pastels to get a fine dust. Whatever colour you use, add a third of black pastel so you can have monochromatic value in order to grade the tones. Almost soak a piece of cotton wool with lighter fluid and then dip in pastel dust. Cover the construction drawing with the lighter fluid and pastel combination using a straight stroke. Do not add any strokes once you have done this.

- Define the light
  source in order to
  determine the sides
  of the product that
  will be shaded (see
  Light and Shadow).
  Use the same colour
  of Pantone marker
  as the background to
  cover the shaded
  sides of the product.
  Use Cool Gray 7
  to draw the product
  shadow on the
  ground.

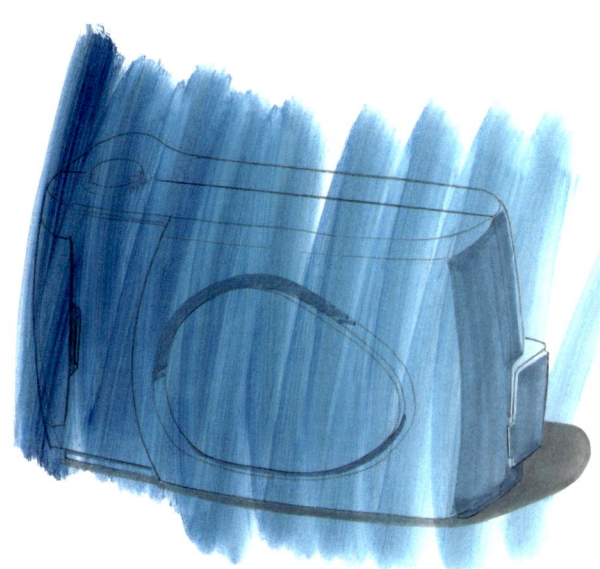

- Use black coloured pencil or a fine liner to draw the construction,
  and also a ruler if needed. Use an eraser stick to create highlights
  and white chalk pencil (see Sketching Materials) for lighter surfaces.

Rubber and mat plastic

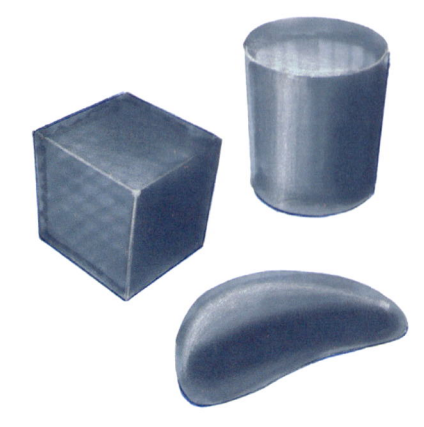

• Apply masking tape around the construction drawing edges. Use a craft knife and cut the tape precisely so that the object is isolated from the background.

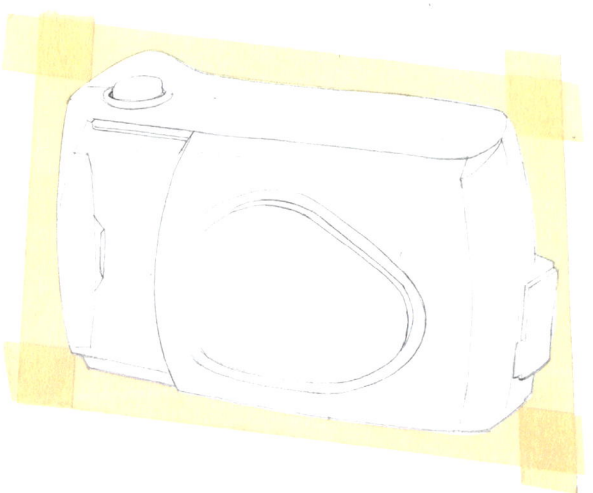

• Use a marker from the colour palette to cover the area of paper surrounded by the tape. Apply one layer with straight strokes.

rubber and platic

- Cover the side of the object with at least two layers of marker to imitate shadow.

- Completely cover se lected object elements with a darker colour. In this case, it is black marker.

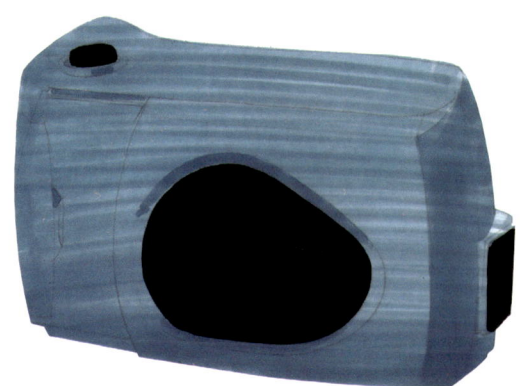

- Redraw the object construction lines with black coloured pencil or fine liners. Use rulers if needed. Then, use a white chalk pencil for high lights and a white coloured pencil to cover lighter surfaces (see Light and Shadow).

Reflective materials

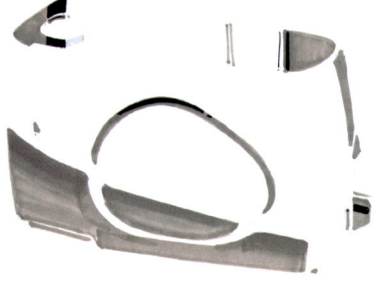

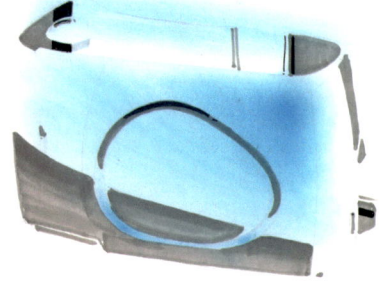

- The key to success when simulating Reflective Materials is planning where to place reflections. It is usual to draw the reflection of the horizon on any rounded surfaces, such as the camera front. Straight lines are used to simulate flat surfaces, such as the camera top.

- Markers are used for simulating high reflections on the product surface. If you want to simulate chrome metal, Cool Gray 7 and black are best to emphasize reflections. For glossy plastic, any colour from the Pantone colour palette will do.

- Pastel dust mixed with talcum powder is used for soft, blurred reflections. Rounded surfaces should be coloured in a gradient from top to bottom. If simulating chrome, it is best to use pure blue mixed with talcum powder to emphasize the sky. If simulating glossy plastic, it is best to use the same colour as the ' previously used marker, mixed with a third of black pastel and talcum powder. Black is very important because of the contrast.

reflective materials

- Redraw the object construction lines, add highlights and erase the pastel around the drawing.

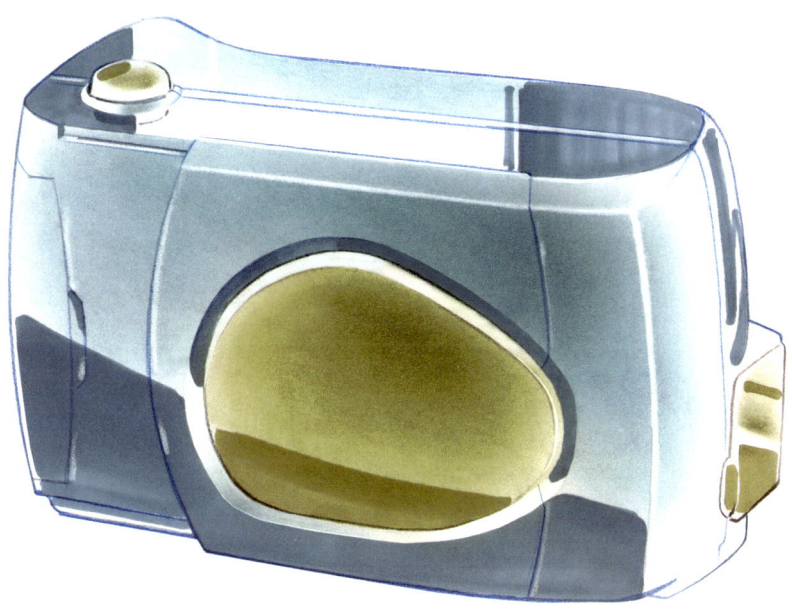

Glossy plastic

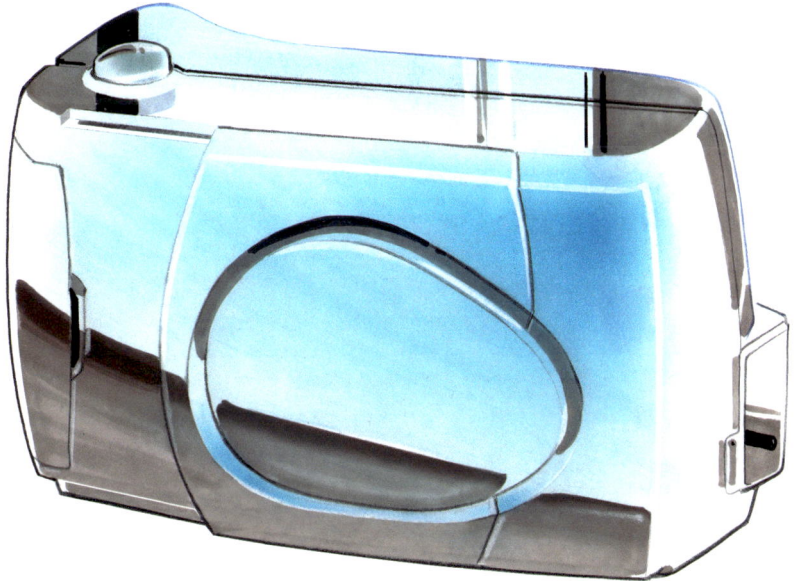

Chrome metal

Brushed metal

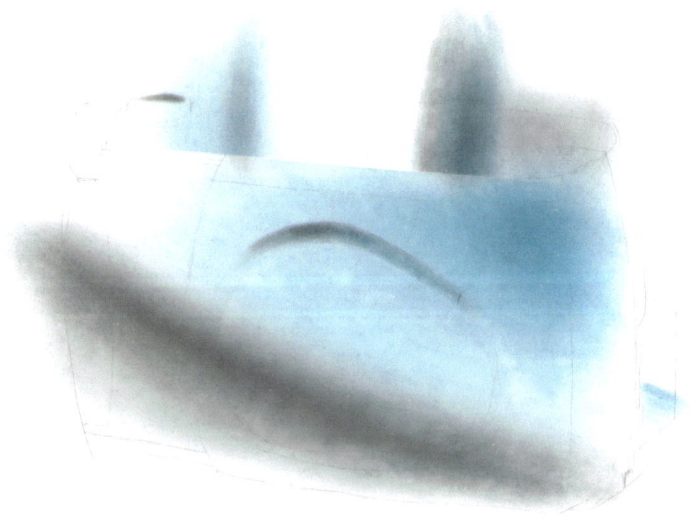

- When simulating Brushed Metal, the same reflection rules are used as for other reflective objects. The only difference is that all of the reflections are soft and blurred because of the surface texture. This is the reason why only pastel dust should be used for brushed metal simulation.

- The front of the camera is a rounded surface, so the reflection of the horizon and background will be visible. Reflections from the sky will show up as a blue gradient from the top right corner. The top surface of the camera will reflect straight lines because it is flat. The straight side of a plain sheet of paper was used as a mask to create the top front edge. Concave surfaces will reflect the ground on the upper surface and the sky on the bottom, this can be seen on the concave surface around the lens cover.

- Redraw the object construction lines with black and blue coloured pencil or fine liner. Outer lines should be thicker than inner lines. Use a ruler if needed.

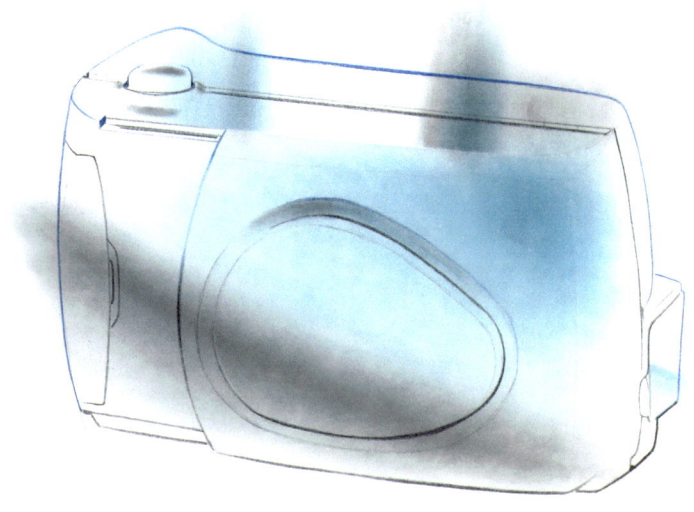

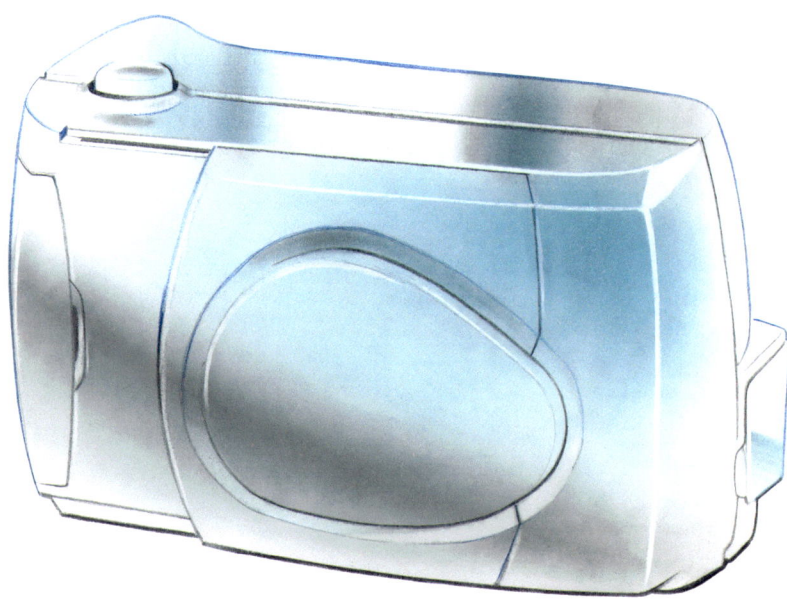

- Add highlights and lighter surfaces with an eraser stick (see Light and Shadow)

Textures

- There are two ways to simulate product textures. One is to imitate texture by drawing with coloured pencils or pastel and cotton wool. The other way is to use the texture of an existing object underneath the paper and use coloured pencil strokes to leave an impression of the surface. Here are examples of how to imitate wood and textured rubber.

- Draw a construction of the product

- Use the same clichés to show the surface lighting. In this case, the wood is matt, so pastel is used to create soft reflections (see sub-chapter Brushed Metal). I wanted to imitate glossy black rubber, so a white stripe of blank paper was left to emphasize a strong reflection (see sub-chapter Reflective Materials)

- Use brown coloured pencil to draw the wood grain. Pay attention to the direction and shape of the grain lines. They will always follow the direction of the object, and will bend and twist with the shape of the path. The rubber part is highlighted from the bottom with white coloured pencil. The surface was imitated by using a finely textured folder.

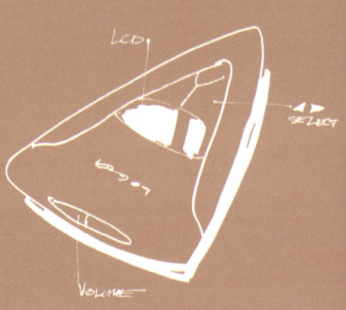

Pitching Sketch

product appeal

product use

the interface

product structure

The Purpose

The Pitching Sketch is a designer's brief of the project in sketch form. It has evolved as a way for designers to communicate with their environment i.e. clients, other departments, engineers, managers and product developers. The main characteristics of this type of sketch are speed, small format with small sketches, and few materials used.

It is used in two ways. Firstly, it is a designer's visualisation response to a discussion within a design group. Also, the designer uses it as a tool to express ideas from other members in the design group; the designer is a communication bridge. It is very effective during meetings and briefings because it facilitates discussion by idea visualization. Once an idea is visualized on the paper, it is much easier to add, modify and manipulate design elements.

The Method

The Pitching Sketch process is principally about the design group meeting, but it also involves a little preparation beforehand. It is not so much about perfect drawing, but about asking the right questions.

The **preparation** consists of gathering useful information about the background of your environment and the future project. It is important to become familiar with the company and subject matter in order to be ready for your first meeting. Here are some tips on research you could perform before this first meeting. These are split into two categories. The first is regarding facts on your client, and the second is about external factors that concern the product i.e. competitors and the market.

Client

- What is the history and background of the company?
- What is their existing portfolio of products?
- What is their technical capacity?

External factors

- What market segments will use your product?
- How will the user perceive the product and what impression will it leave?
- Either research the functional needs in the market for your product, or look at existing products as a basis for redesign
- What characteristics of the product do competitors focus on e.g. design, technology, environment?
- What is the current social context of the product and how will this change in the future?

In the **design group meeting**, you are concerned with finding out about four design aspects. These four aspects are:

Product Appeal; how does the product appeal to the user?
Product Use; how will the product behave?
The Interface; how will the product communicate with the user?
Product Structure; how will the product materialise?

Based on these questions, you can formulate concepts. These concepts will then be based on the demands of your environment and you will have defined the constraints of the design problem.

A design group meeting will demand a series of Pitching Sketches. These sketches will define the different aspects of the product. However, not all of the aspect concepts will fit together to make a coherent design. Instead, they give the overall brief of the project. So, here we see that sketching is not only about drawing, but actually becomes a method to facilitate the design process. The Pitching Sketch is acting as a set of guidelines for the designer and as a set of questions for the design group.

Due to discussions within a design group meeting moving rapidly, the designer will not find the time to draw proper constructions of the product. Instead, you will have to draw a simple caricature that everyone within the group will understand. Therefore, the Pitching Sketch should use the Outline Drawing style as discussed in the Sketch Elements chapter.

Design Aspect: Product Appeal

-Draw a Perspective view of the product. Add some more drawings from different viewing positions if necessary.

    Questions you should ask the design group:
• Who is the user of our product?
• What is the semiotic term that describes our product?
• How could the product possibly be branded?
• What are the external design elements e.g. buttons, displays, wires?

-Once the answers are obtained, try to:
• Choose the shape orientation: horizontal, vertical.
• Choose the shape style: amorphic, cylindrical or cubical.
• Define the external elements of the product.
• Generate ideas of what the interface will look like; a shape description only.

Figures: **projection, perspective**
Styles: **outline style**

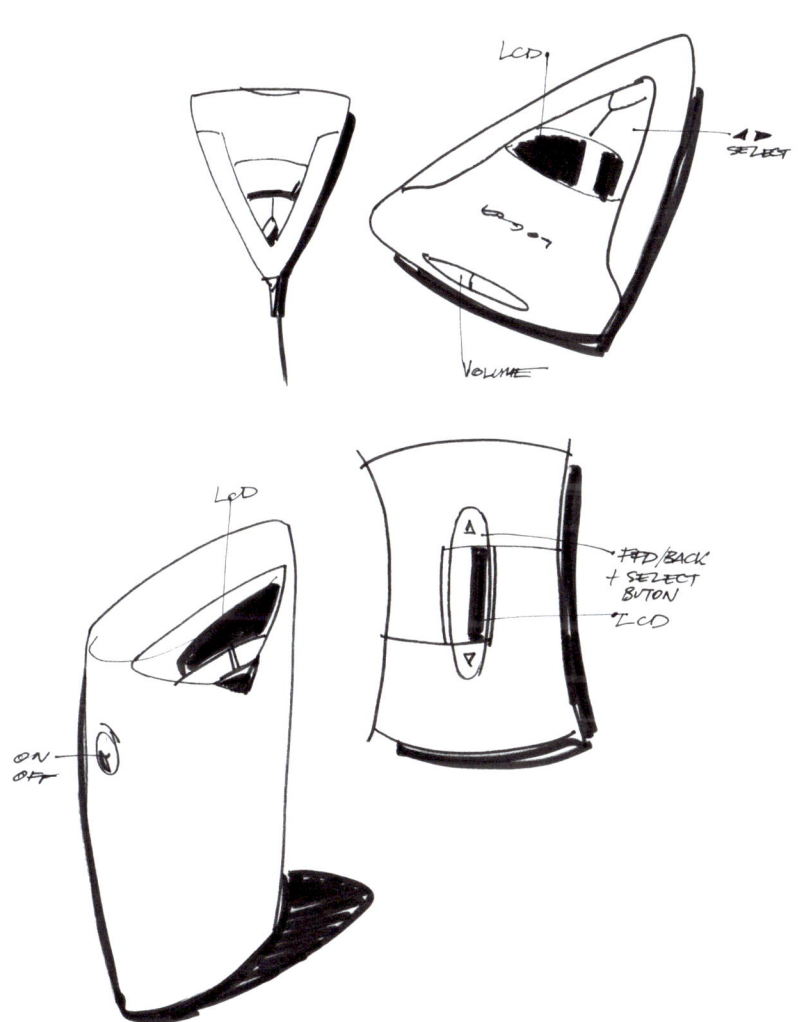

LCD

SELECT

Volume

LCD

FFD/BACK
+ SELECT
BUTON

LCD

ON
OFF

Design Aspect: Product Use

-Draw the Timeline view to explain how the product functions and how it will be used.

-Questions you should ask the design group:

- What are the functions of our product?
- What is the behaviour that fulfils these functions?
- What does the behaviour consist of e.g. preparing, use and not-in-use modes?

Figures: **timeline, projection**
Styles: **outline style**

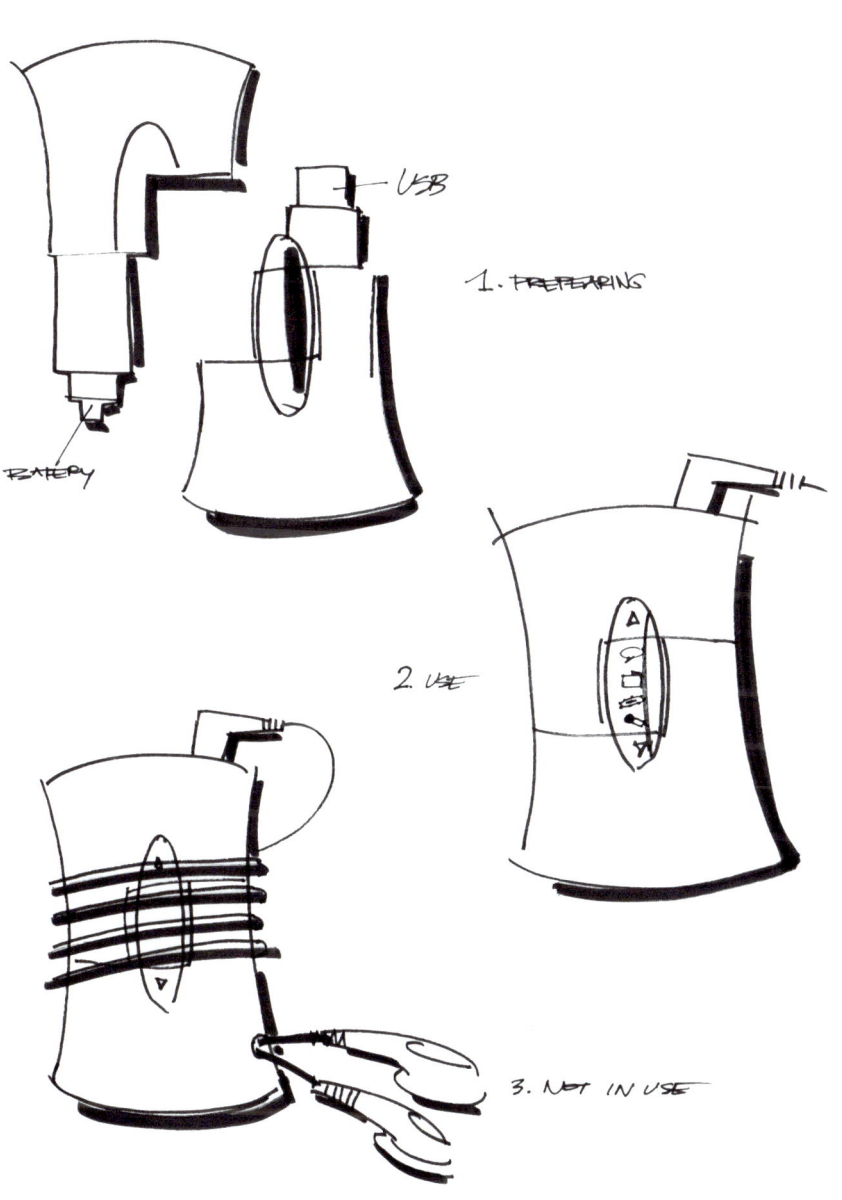

USB

1. PREPEARING

BATERY

2. USE

3. NOT IN USE

Design Aspect: Product Interface

-Draw an interface Star view that explains the how the interface will be used.

-Questions you should ask the design group:

• What is the interface behaviour that will fulfil the functions defined by the Product Use sketches?
• What does the interface consist of?
• Is there any need for additional graphic design?

Figures: **the star**
Styles: **outline style**

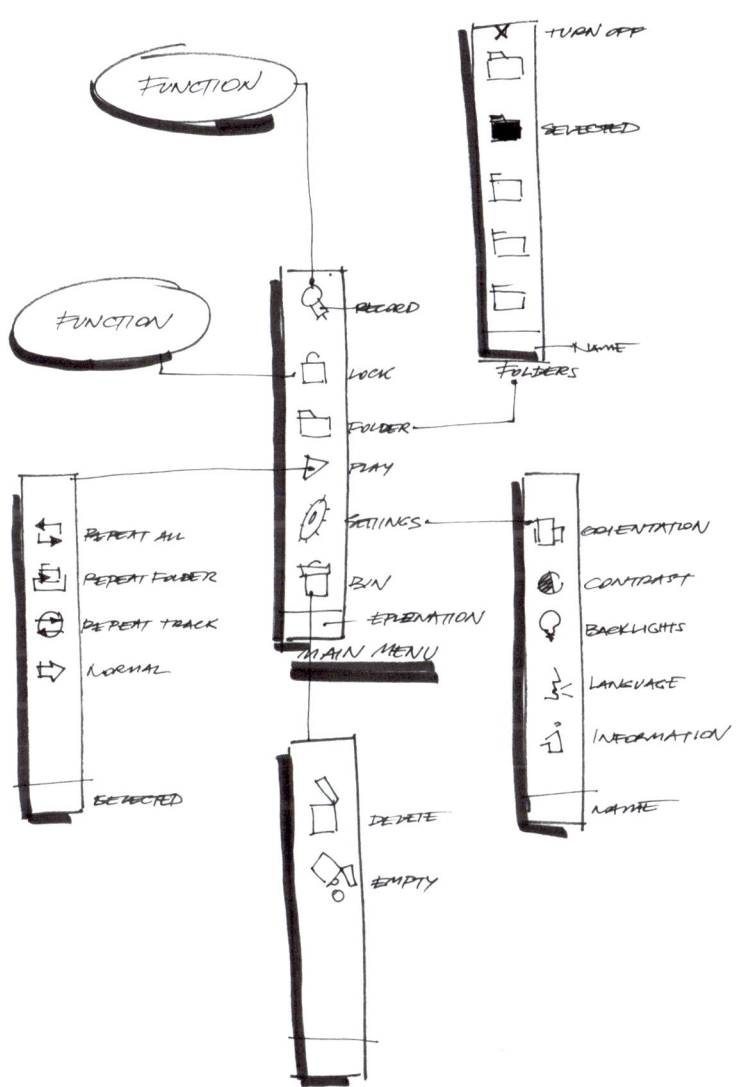

FUNCTION

FUNCTION

TURN OFF

SELECTED

NAME

FOLDERS

RECORD

LOCK

FOLDER

PLAY

SETTINGS

BIN

EXPLANATION

MAIN MENU

REPEAT ALL

REPEAT FOLDER

REPEAT TRACK

NORMAL

SELECTED

ORIENTATION

CONTRAST

BACKLIGHTS

LANGUAGE

INFORMATION

NAME

DELETE

EMPTY

Design Aspect: Product Structure

-Draw the inner and outer structure of the product

-Questions you should ask the design group:

- What type of structure does the product have and how do the parts fit together?
- What materials do you intend to use?
- Which production methods will you use?

Figures: **ghosting, explosion, sections, detail icon**
Styles: **outline style**

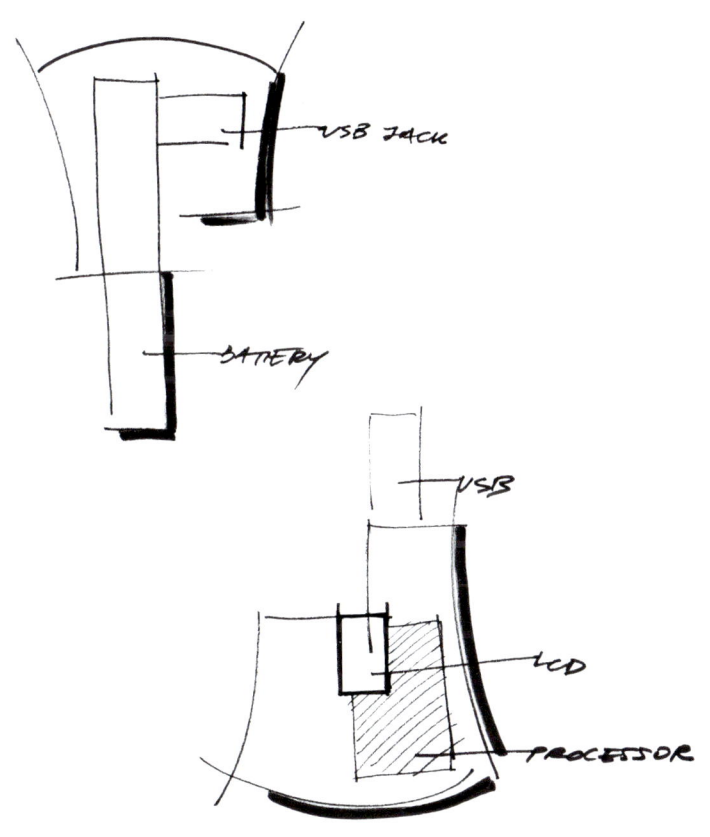

USB JACK

BATTERY

USB

LCD

PROCESSOR

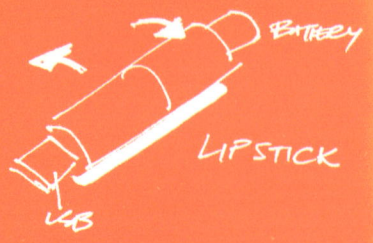

BATTERY

LIPSTICK

USB

Memo Sketch

gathering information

taking notes

documenting ideas

using cutouts

The Purpose

The Memo Sketch has evolved as a technique for industrial designers to gather and memorise information about ideas that can be developed. This technique also allows the designer to form 'provocations' about design aspects and release their creativity. Due to this, Memo Sketching ideally suits the designer's personality by being intuitive and perceptive, and starts the important process of design thinking on an unconscious level. The resulting Memo Sketches are very suitable for the Concept Sketching phase and therefore, this technique is the industrial designer's way to research.

The main characteristics of this type of sketch are speed, small format with small sketches, and few materials used. The designer draws or constructs the shape, and uses text or specific Views in order to describe design aspects. This is done with a quality of sketch that he will be able to recognise his idea at a later date. The Memo Sketch is not only used in the design studio, but also in your daily routine i.e. whilst travelling, in a café or in bed. Memo Sketching is not a typical design activity, but more a design habit.

The Method

-For a memo sketch these are the topics to focus on:

- Definition of the goals the product has to achieve.
- Definition of the user group and user profile.
- Definition of the social context and use of the product.
- Definition of the need for certain functions among users.
- Definition of the identity of the product and the elements that create a fulfilling user experience (the 'X-factor').

-For a Memo sketch, you can base your provocations on these parameters:

- Imagine how the product could look in the future. Will the need still exist at that point in time? What other functions could satisfy that need?

- Create new user needs. Try to predict the future needs of certain user groups according to their lifestyle and habits? Imagine users of your product; their age, background and lifestyle.

- Use aesthetical and behavioural properties of the product to create a user experience. Try to enrich this experience with non-functional details such as smell, colour, materials, texture, and visual identity. Discovery of clever, creative functions will also significantly enhance the experience. Try to grab the user's attention by making them relate the product to the company ethos.

- Use aesthetical and behavioural properties to fit the product identity with the client identity.

- Define possible added values for the product e.g. scroll buttons on a mouse

- Define the production complexity and materials according to the budget.

- Define product lifecycle, from production to disposal, and even recycling if possible.

In order to make a Memo Sketch, the choice of materials is down to the individual. Some designers like to use fine liners because it forces one to describe shape, process or technical details more precisely. Others use coloured pencils very successfully by keeping their drawing two dimensional.

When using a memo sketch it is important that each sketch focuses on only one aspect of the design. This reduces misunderstandings and makes the sketches easier to read later. Memo sketches should not be too detailed but can be of an iconic style, one projection or ghosting will do. The most important thing about memo sketches is that they are recognizable. Drawing style and quality is not important for this type of sketching.

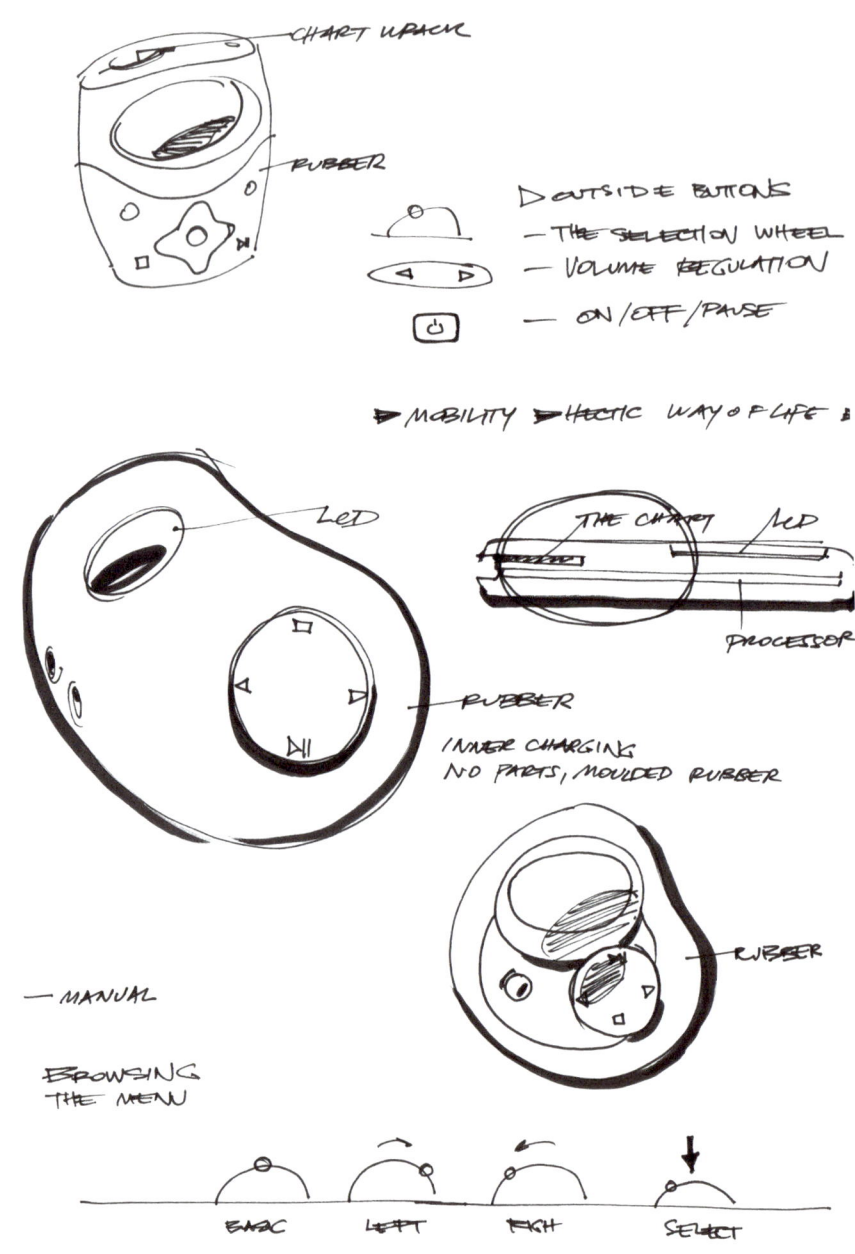

CHART WRACK

RUBBER

▷ OUTSIDE BUTTONS
— THE SELECTION WHEEL
— VOLUME REGULATION
— ON/OFF/PAUSE

▶ MOBILITY ▶ HECTIC WAY OF LIFE ▮

LED

THE CHART   LCD

PROCESSOR

RUBBER

INNER CHARGING
NO PARTS, MOULDED RUBBER

RUBBER

— MANUAL

BROWSING
THE MENU

| BACK | LEFT | RIGH | SELECT |

• Gathering relevant information from the internet,
  literature and photography ('brain food').

• Taking notes and sketching during meetings.

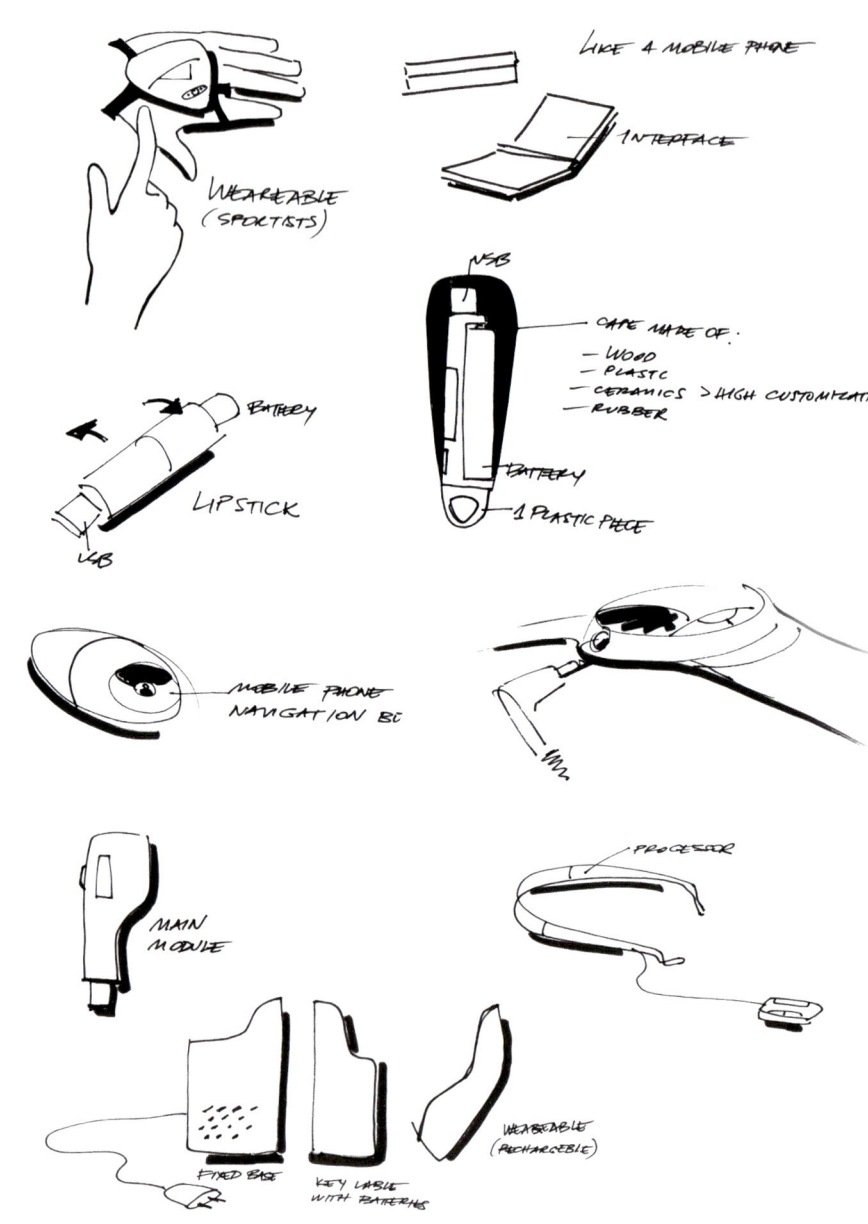

- Documenting sudden ideas (brainwaves) that could
  be used and refined in the Concept Sketching phase.

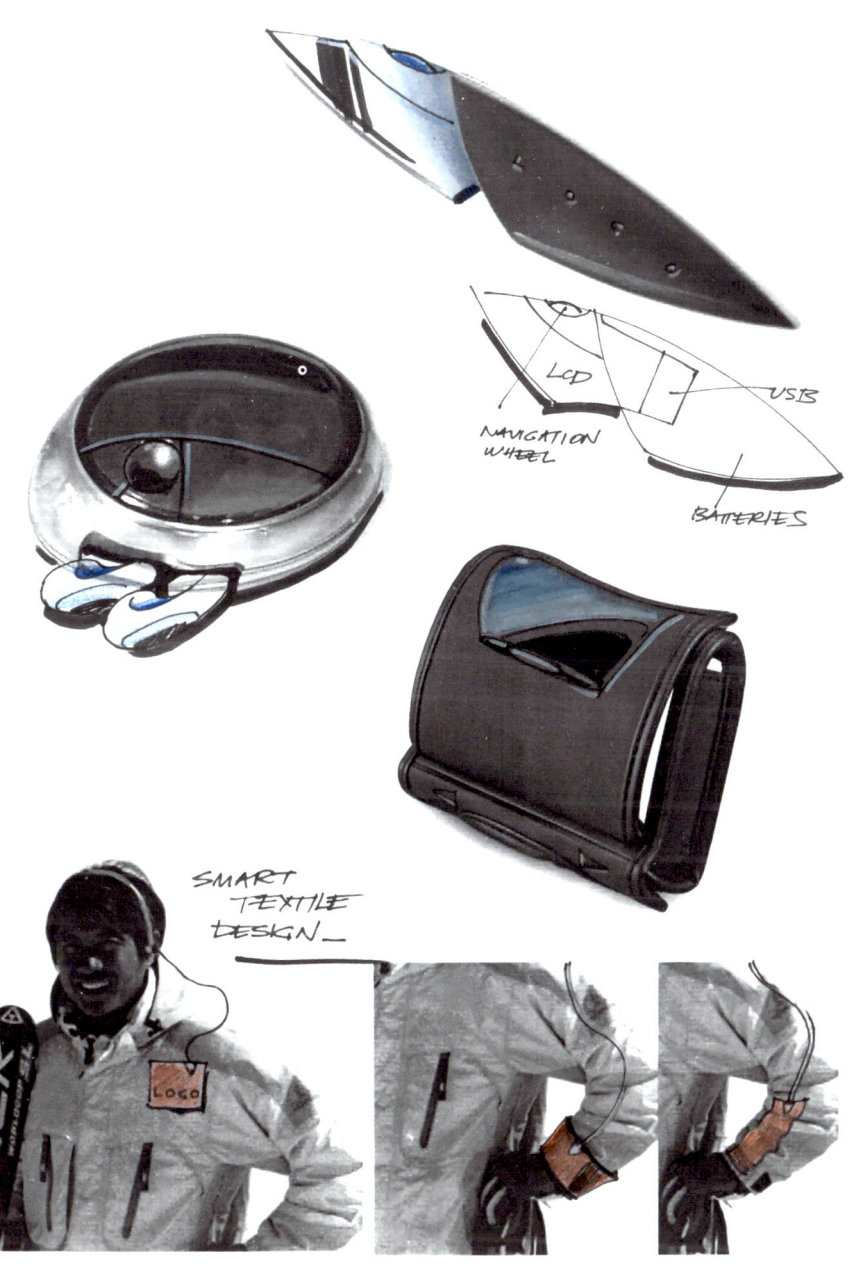

LCD

USB

NAVIGATION
WHEEL

BATTERIES

SMART
TEXTILE
DESIGN

LOGO

- Adding notes and sketches on collected material such as photo
  copies, printed images or even text from selected sources e.g. the
  internet, magazine cut-outs etc.

Concept sketch

product function

shape of the product

the interface

product parts

The Purpose

Designers traditionally use Concept Sketches to express themselves. They use it to understand problems and investigate a broad diversity of solutions, through the process of visualization and redrawing (refinement). Concept Sketching is a tool to record and support creative thinking. It is also a facilitator because it enables spontaneity and creates the possibility to change direction in the design process.

The purpose of a Concept Sketch is to generate ideas from provocations, give the product a shape, decide upon production systems, and explain functions. It is a very complex activity but by mastering it, you are enabled to communicate with the design problem through drawing.

The Method

The Concept Sketch is used to define what you, the client and the user want from the product. Here are some things to think about when planning the process:

- Establish your personal goals for the product according to the client's own goals, suggestions and capacities.
- Research the market for existing products, and establish a communication network with institutions or individuals connected with the product development area. Communication with distributors could prove useful in a redesign.
- Plan feedback on your concepts through market research and client opinions.

A Concept Sketch usually only provides limited information on product materials or textures. It should not represent a realistic image of the product, but rather explain design aspects of the product. Very often, Concept Sketches are used in combination with rough physical models e.g. clay, foam or wood.

A Concept Sketch should describe the four different Design Aspects that demand attention during the design process. Please note that the aspects presented in this and the Pitching Sketch chapter are the same, but given in a different priority order. This is because Concept Sketches have a different design focus. However, the priority order is not set in stone, but just a reference in order to help your work.

Design Aspect: Product Use

-Define the use of the product:
- Concentrate on user needs. What are the functions
  that could fulfil these needs?
- Concentrate on the functions. What are the behaviours
  that could fulfil these functions?
- How could the behaviour be developed to satisfy the user experience?

-Decisions:
- Define all the possible scenarios of product use.
- Define the functions in each of the scenarios.
- Describe the behaviour that fulfils these functions.
- Describe the parts of product that fulfils the function.
- Describe the scenario in which the product will be used.

-Technique:
- Use a propelling pencil to draw a Perspective View of the product
- Name all the functions of the product and mark them on the
  perspective drawing
- Use a Timeline to explain how the product functions and how it will
  be used. The most desirable views to use are Profile or Perspective.
  Use a propelling pencil for the construction drawing.
- Finish it in Free Style or Shading

Figures: **timeline**
Styles: **free style, shading**

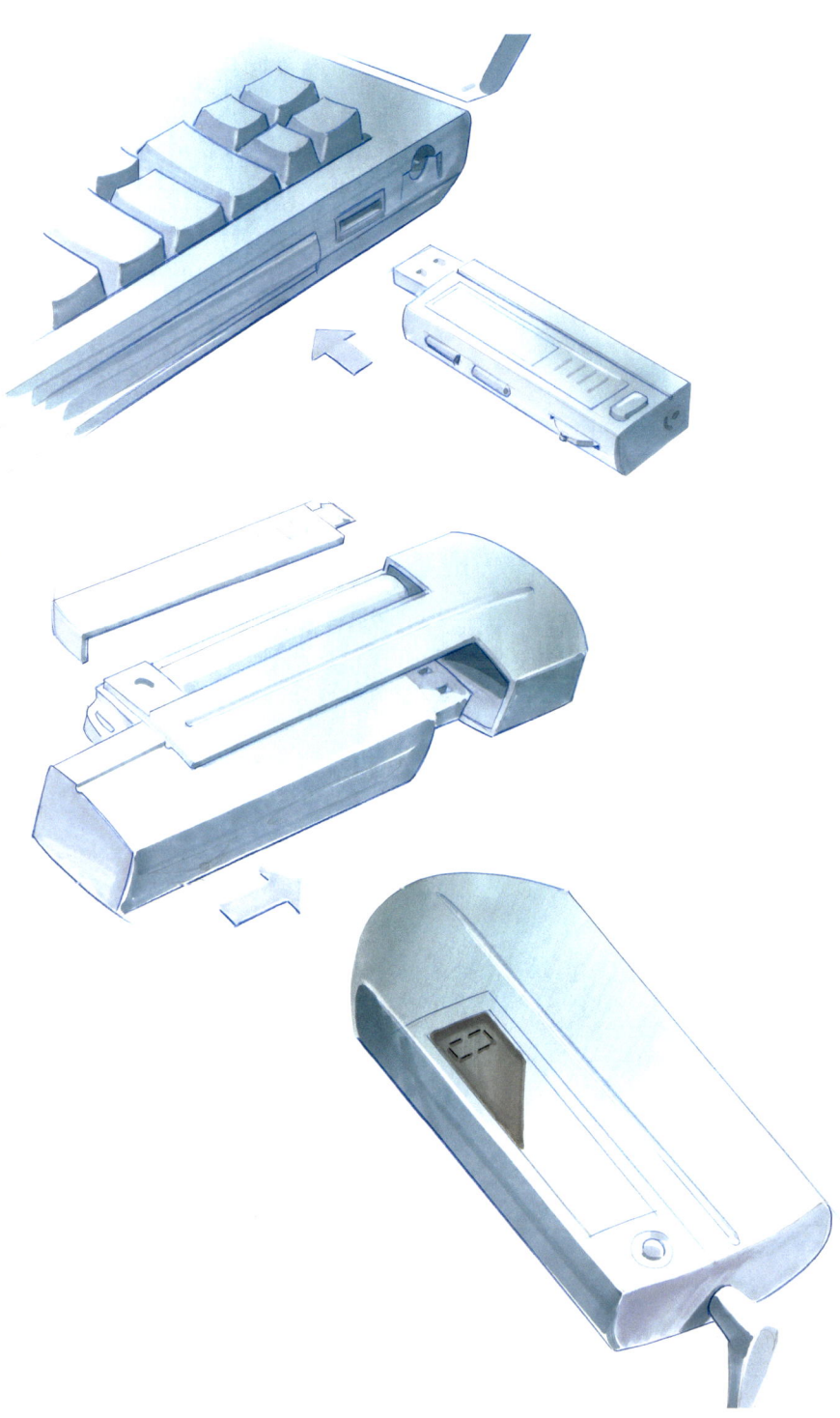

Design Aspect: Product Appeal

-Use collected research material to define your Product Appeal:
• What is the user profile for this product?
• What is the semiotic term that describes the product?
• What does that semiotic term symbolize?
• What is the size and proportion of the product?
• What type of construction does it use, e.g. shell construction, tubes, mounted flat panels, textiles

-Decisions
• Choose the shape orientation: horizontal or vertical
• Choose the shape style: Amorphic, Revolved, Cubic, or a hybrid.
• Define the external elements of the product.
• Describe what the interface will look like, a rough description only.
• Illustrate joints between the parts of your product

-Technique
• Use marker paper and a propelling pencil to draw a series of sketches that explain the object shape and the external appeal of the product. Use Perspective and Projection Views.
• Choose the most successful design of the previous step and draw it to a larger scale. If necessary, sketch several different Views of the product. Make a new Perspective drawing to describe the product if it changes shape during use.
• Use the Net View and Shading style to describe the form of the object

Figures: **perspective, netting perspective**
Styles: **free style, lightening, shading, materialization**

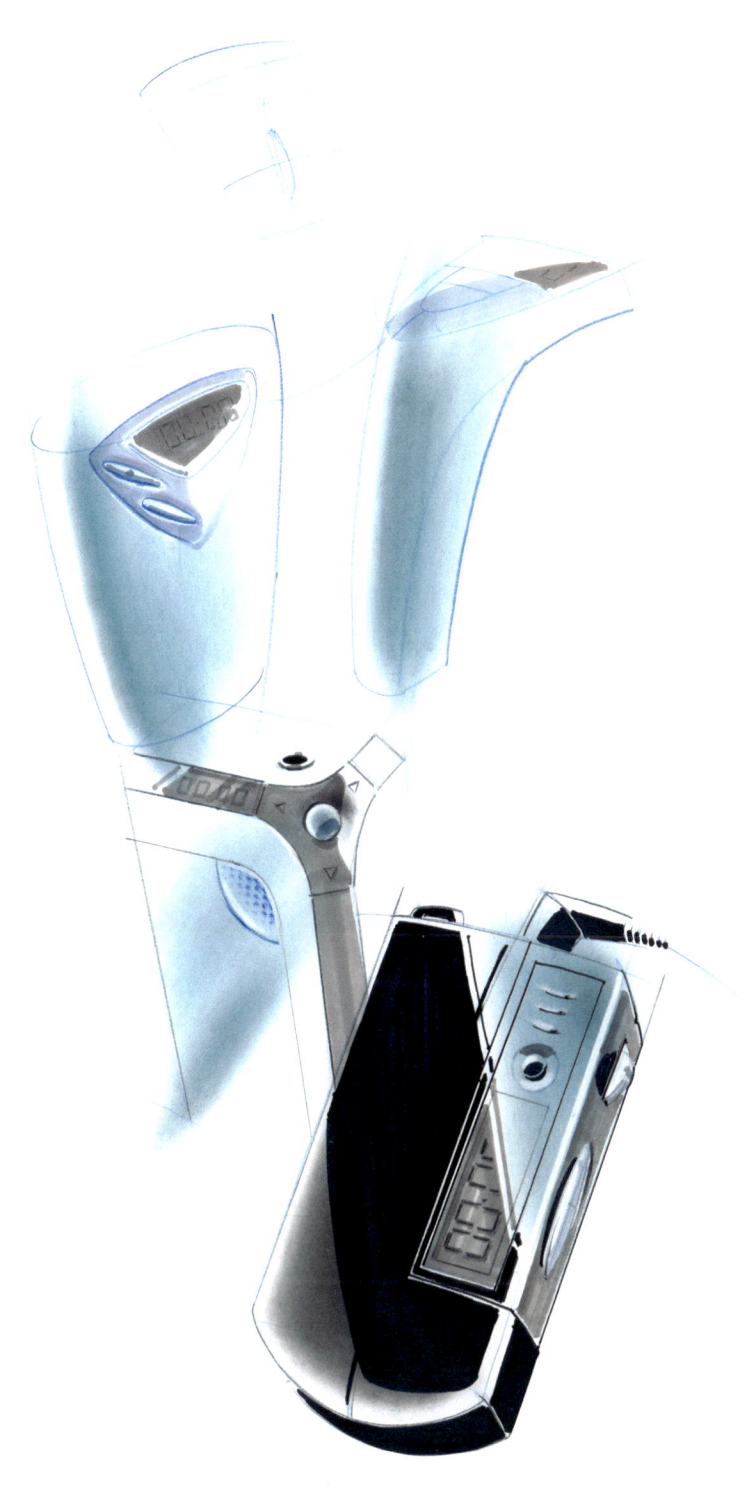

Design Aspect: Product Interface

-Use collected research material to define your Product Interface:
• How to control the functions?
• What is the behaviour that fulfils these functions?
• What does the interface consist of?
• Does the interface require any graphic design?

-Decisions to make:
• Define the function and behaviour of the interface
• Define the interface elements
• Define the external appeal
• Define the graphics

-Technique:
• Use marker paper and a propelling pencil to draw the interface shape
  and elements with feint lines. Use a Projection View.
• Choose the most successful design of the previous step and draw it to a
  larger scale
• Draw a Star View to explain 'if and then' functions of the interface, and
  render in a Highlight or Shading Style.

Figures: **the star**
Styles: **lightening, shading**

TRACK ONE / REPEAT FOLDER / REPEAT ALL / A→B / NORMAL / SHUFFLE FOLDER.

SELECTION NAME

POP / CUSTOM / NORMAL

SELECTION NAME

FIN / EMPTY

SELECTION NAME

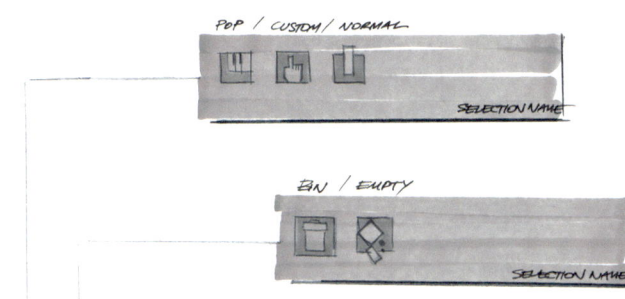

PLAY / EQ / DELETE / SETTINGS / RECORD / LOCK.

SELECTION NAME

PLAYING MODE

00 · 00 · 00   FOLDER NAME

NAME AUTHOR · ALBUM'S NAME

MOVING →

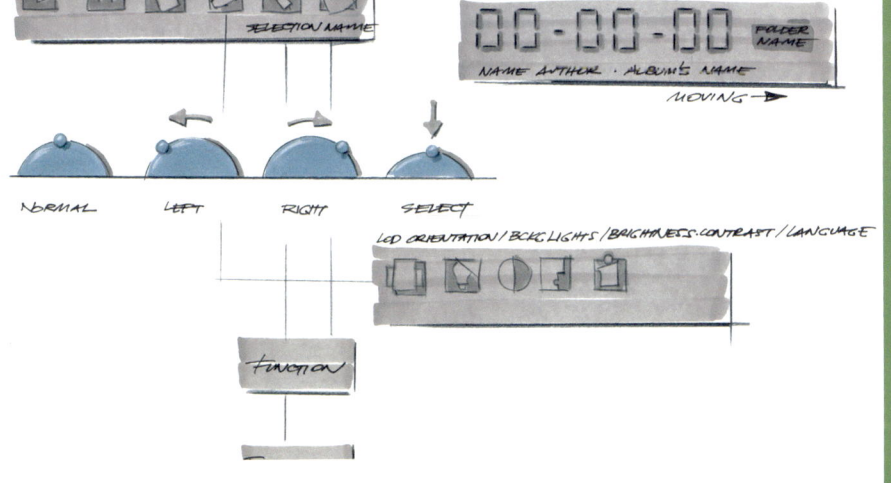

NORMAL   LEFT   RIGHT   SELECT

LCD ORIENTATION / BACK LIGHTS / BRIGHTNESS · CONTRAST / LANGUAGE

FUNCTION

Design Aspect: Product Structure

-Use collected research material to define your parts, assembly and materials:
• What structure and parts does the product have?
• What materials are the parts made from?
• Which production methods will be used?
• What is the most effective production method for a certain material or shape?

-Decisions to make:
• Define the number of parts and name them.
• Define the material of each part and possible production methods.
• Define the joints and possible ways of assembly.
• Make sure the shape, material choice and production methods complement each other.
• If the product has moving parts during use, then define the mechanisms that enable the movement.

-Technique
• Use marker paper and a propelling pencil to draw the product parts with feint lines. Use the Explosion View to describe parts and add Detail Views to explain joints and mechanisms. If the product shape is too complicated, use an Explosion View combined with a Projection View to focus on the idea.
• Choose the most successful design of the previous step and draw it to a larger scale
• Draw an Explosion View to explain the product parts in a Highlight or Shading Style.

Figures: **explosion, sections, detail icons**
Styles: **free style, shading**

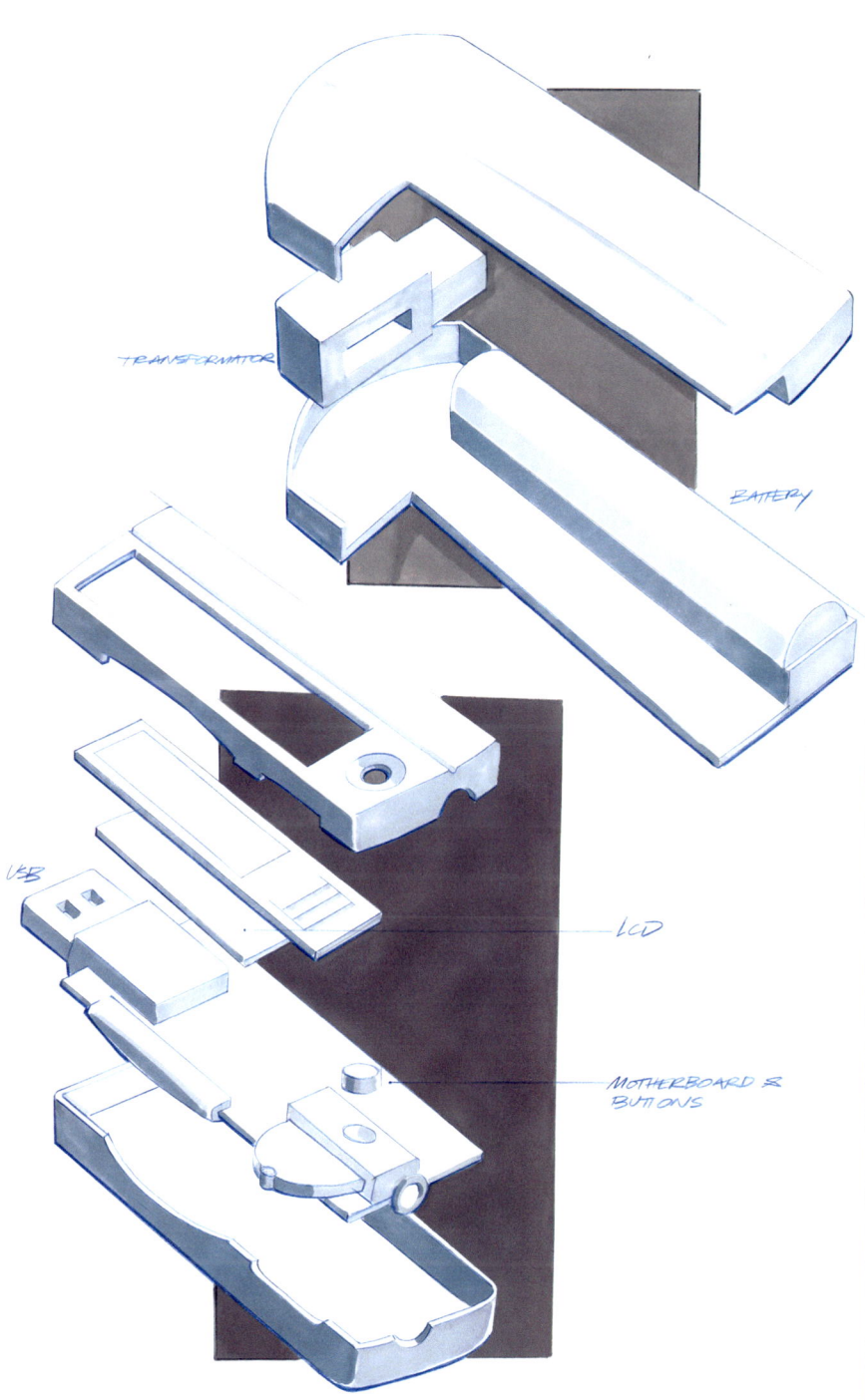

TRANSFORMATOR

BATTERY

USB

LCD

MOTHERBOARD &
BUTTONS

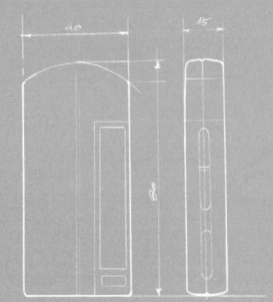

Specification sketch

product shape

proportion and detail

model drawing

The purpose

A Specification Sketch uses technical drawings and precise renderings to specify different aspects of the product and will prepare you for the physical or CAD modelling stage. It is a process of synthesis, which revises all the previous concepts and evaluates the suitability of the product Design Aspects. The Specification Sketch will allow you to see if it is possible to implement the concept in three dimensions.

A Specification Sketch puts some physical boundaries on the concept idea and the Design Aspects. This phase is used to revise the concepts in order to make it physically possible to produce them.

The Method

-In the Specification Sketch phase, there are five things to concentrate on:

- Choose a concept
- Design a hybrid concept
- Product Shape
- Proportion and Detail
- Model Plan

Choose a Concept

At this stage, it is time to evaluate all previous decisions and try to find out what concept will fulfil the demands of the task. It is a summary of all the good and bad sides of your design. It is important to not only use common sense, but also intuition in order to consider all the product factors, especially those from the client and product developer. Also, it is good to ascertain the state of the market.

However, there are many methods for design evaluation. The intention of this book was not to describe these methods. Instead, the intention was to propose a possible system for you to establish your personal values, and make your own choice from a designer's viewpoint.

Here are some tips to make your choice:

- Sort your concept sketches by dividing them into groups and give them reference names. Divide the concept names into groups and subgroups on a sheet of plain paper in order to get a better view of the situation.
- If it is not possible to choose a concept by simple elimination, use a matrix ranking method with the four Design Aspects described in the Concept and Pitching Sketch chapters:

- Function and behaviour of the product
- Product shape and appeal
- Interface and usability
- Parts, assembly, production and materials

A) Use a ranking of 1-4 to describe the Design Aspect importance
B) Rank the concepts against each design aspect using 1-5.
C) Multiply together the concept rating and the Design Aspect importance.
D) Add together these multiples for each concept to get a total rating score.

Design the Hybrid Concept

- At this stage, it is still possible to generate new ideas. Try designing a Hybrid Concept by combining existing concepts to create a better solution. Be careful not to destroy the core idea.
- Use tips from the Concept Sketch chapter to design and describe the Hybrid Concept.
- Place marker paper on the existing drawings and use it as a source for new drawings

Product Shape

- Draw rectangles to represent the product dimensions. These will act as a reference for the Projection View. Use a propelling pencil and plain white paper to create this plan.

- Draw the product shapes within the reference rectangles.

- Refine the shape and dimensions so that the product keeps its character. Define dimensional boundaries so that you know how much you can change them.

- Use a Ghosting View and draw reference rectangles to represent the inner elements of the product. Sketch each of them with different coloured pencils. Try to fit them into the product shape.

- Mark the dimensions of the product, including all its inner and outer elements.

- Use a new sheet of marker paper on top of the plan, and redraw the projections. Draw only the external elements, ignoring Ghosting.

- Render the drawing in a Highlighting or Shading Style.

Figures: **projection, ghosting**
Styles: **outline, shading, lightening**

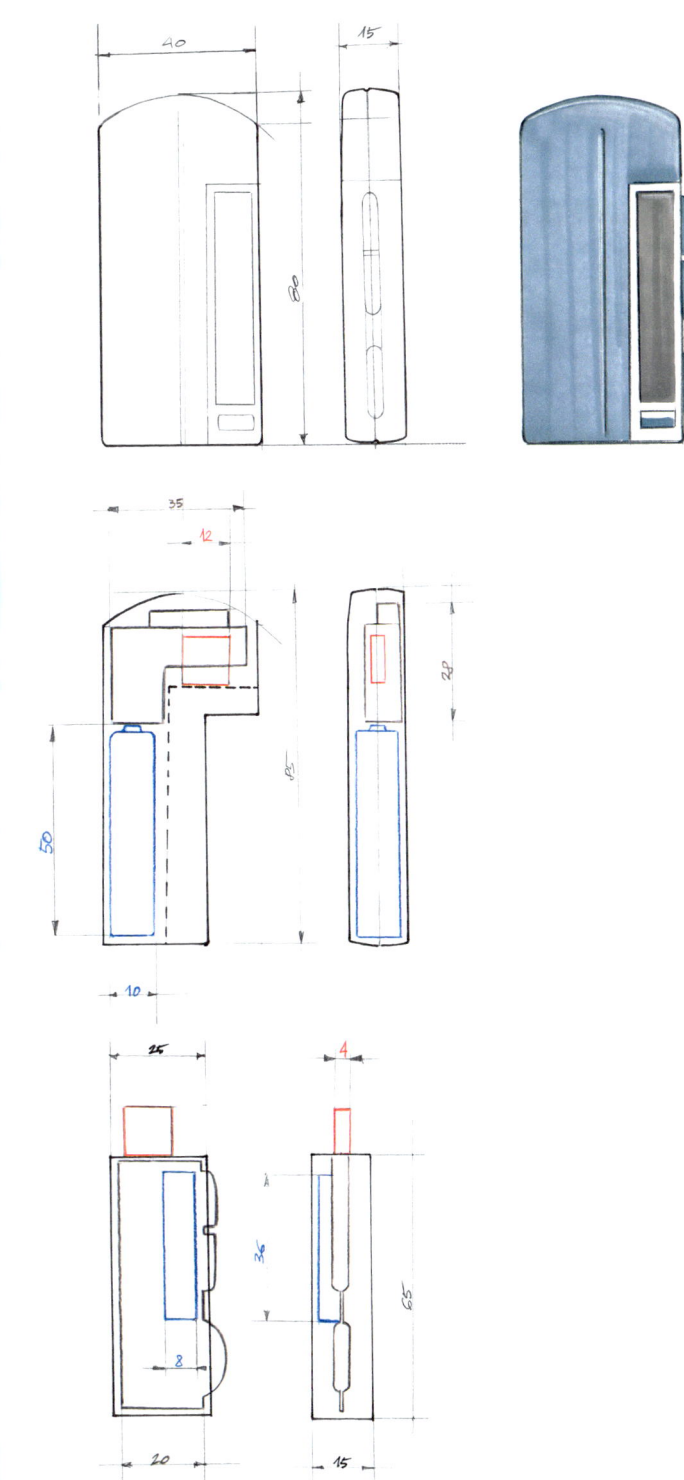

Proportion and Detail

• Use a propelling pencil and plain paper to draw at least two different Perspective Views of the product. Take care that the proportions are as accurate as possible. This drawing will be used as a plan for the Proportion and Detail study.

• Use a new sheet of marker paper to place on top of the plan and trace the sketches in pencil. Redesign the shape within the dimensional boundaries you defined earlier. Also, redesign details and the interface to better fit the Product Appeal. At this stage, we want to see how the product will appear when manufactured from various different materials. In order to explore this, you should render many sketches of your product using different material styles presented in the Styles chapter e.g. Brushed Metal, Reflective Materials or Textures.

• Repeat this procedure until the result is satisfactory.

Figures: **perspective**
Styles: **materialization**

Model Drawing

- Use plain paper of necessary size and the propelling pencil to draw Projection Views of the product. Try to keep it in proportion 1:1 if possible.

- List the parts and create a step by step plan for a future CAD or workshop model. In most of the cases these parts will be redesigned for a 'solid' manufacturing. Only in the case of prototype production the shell will be necessary.

- State all the dimensions and show every detail.

- If the product is amorphic and has more difficult shape, it is useful to draw a Net View and use it as a reference for modelling.

Figures: **projection**
Styles: **outline, shading**

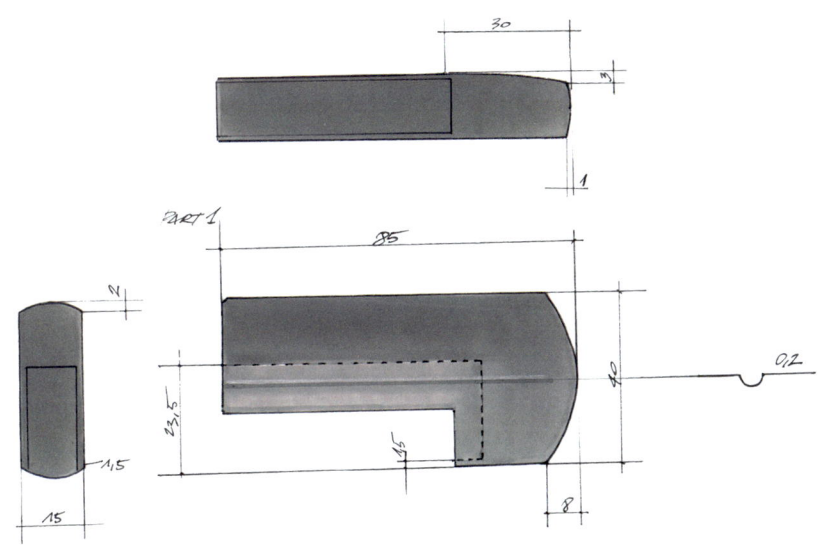

30

3

1

PART 1

85

2

20

0,2

23,5

1,5

5

8

15

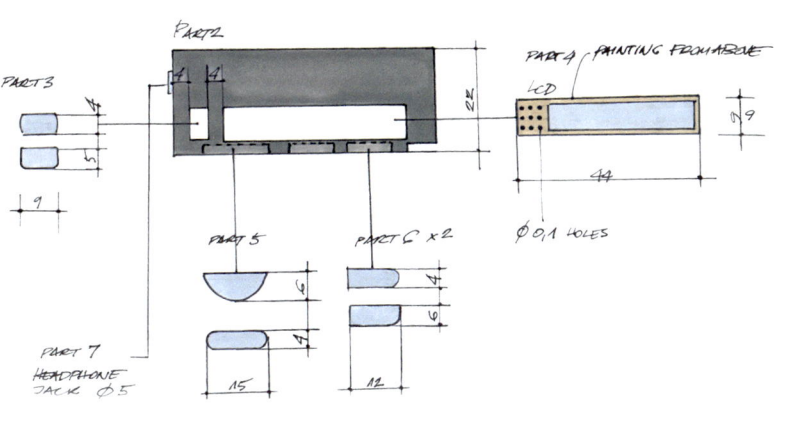

PART2

PART3

PART 4 — PAINTING FROM ABOVE

LCD

1

1

2,9

22

5

44

9

ø 0,1 HOLES

PART 5

PART 6 x 2

6

4

PART 7
HEADPHONE
JACK ø 5

4

6

15

12

What's next?

This is the end of the guide; however it is only just the beginning. Now your ideas are ready to be transformed into a CAD or workshop model. All of the sketches you have been drawing will be very useful in this stage.

When creating a **CAD model**, pay attention that the software does not ruin your form or main idea. Scan your Model Drawing or Product Shape Specification sketch. Use these scans as a background in the projection windows of your CAD program in order to accurately trace the shape. This will reduce the chance of losing your form when converting the design into three dimensions. Your design, sketch precision and desired output will dictate the choice of 3D software. Solid based programs such as Inventor, Pro Engineer or Catia will work better for Revolved and Cubic shapes, while surface based programs such as Rhino, Studio Tools or 3ds max are more suited to Amorphic shapes. If you need a CAD model for design of production tools such as moulds, solid based software is preferable.

When creating a **workshop model**, pay attention to your parts and production plan. A realistic timetable and plan is crucial for success. Do not allow the choice of workshop tools or materials to simplify or destroy the core design idea.
Once again, the choice of material will be determined by the nature of your design, sketch precision and the purpose of the model. Solid materials such as foam or wood will fit Revolved and Cubic Shapes better, whilst clay material will work better with Amorphic Shapes. To keep the proportions and character of your shape, use Projection Views from the Model Drawing. For some more complicated shapes, it is good to have templates so that you can trace the parts onto the material. If you deal with Amorphic shapes in clay material, it is good to have templates for the characteristic sections, such as the top and front profiles.

Notes

Notes

Notes

Notes

Notes

In Closing

As it was said at the beginning, design is a skill and it demands practice. This guide presents a very easy way for you to learn how to use sketches for designing. Therefore, I recommend that you use it when working on design projects.

However, this guide is not a tutorial book and you should not aim to copy the drawings that have been presented. Instead, you should use the methods presented in your own way.

It will take some time before your sketches reach a satisfying level. Do not worry if the sketching style that you are trying to achieve does not work first time. Try to focus on describing the shape and concept instead.

I hope it has helped you to learn not only how to sketch, but also how to design by sketching. I also hope that this will be a useful reference for you in the future and a challenge for you to improve. Enjoy your work!

Reference List

Lawson B, 1997: How designers think;
the design process demystified, London.

De Bono, E, 1996: Serious creativity, London

Durling D, Virtual personalities,
http://www.dmu.ac.uk/ln/4dd/synd4e.html

Ulrich Karl T., Eppinger Steven D.
Product Design and Development